Lecture Notes in Electrical Engineering

Volume 267

For further volumes:
http://www.springer.com/series/7818

Guangyu Sun

Exploring Memory Hierarchy Design with Emerging Memory Technologies

 Springer

Guangyu Sun
Peking University
Beijing
People's Republic of China

ISSN 1876-1100   ISSN 1876-1119   (electronic)
ISBN 978-3-319-00680-2   ISBN 978-3-319-00681-9   (eBook)
DOI 10.1007/978-3-319-00681-9
Springer Cham Heidelberg New York Dordrecht London

Library of Congress Control Number: 2013943950

Printed on acid-free paper

Springer is part of Springer Science+Business Media (www.springer.com)

Contents

Chapter 1
Introduction

1.1 Motivation

In this section, we introduce several challenges, by which the leverage of emerging technologies are motivated. From the architectural view, we first discuss the increasing requirement of on-chip memory and bandwidth. Then, the issues of increasing power consumption and the vulnerability to soft errors are addressed, especially for processors with large integrated memory.

1.1.1 Requirements of Large Memory and High Bandwidth

The diminishing returns from increasing clock frequency and exploiting instruction level parallelism in a single processor have led to the advent of Chip Multiprocessors (CMPs), which are processing cores forming a decentralized micro-architecture that scales more efficiently with increased integration densities [1, 2]. The integration of multiple cores on a single CMP is expected to accentuate the already daunting "memory wall" problem [1, 3]. Supplying enough data to a chip with a massive number of on-die cores will become one of major challenges for performance scalability.

The improvements in microprocessor also result in the requirement for high bandwidth access to the memory [3]. Especially for the CMPs, when multiple threads are running at the same time, both the capacity of caches and the bandwidth to main memory need to be increased in order to improve the performance. Figure 1.1 shows the hypothetical case where the number of threads, or separate computing processes running in parallel on a given microprocessor chip has been doubled [4]. To hold the miss rate constant, it has been observed that the amount of data made available to the chip must be increased [4]. Otherwise, it makes no sense to increase the number of threads because they will encounter more misses and any potential advantage will be lost. As shown in the bottom left of Fig. 1.1, a second cache may be added for the new thread, requiring a doubling of the bandwidth. Alternatively, if the bandwidth is

G. Sun, *Exploring Memory Hierarchy Design with Emerging Memory Technologies*, Lecture Notes in Electrical Engineering 267, DOI: 10.1007/978-3-319-00681-9_1, © Springer International Publishing Switzerland 2014

Fig. 1.1 When the number
of threads doubles, if the total
bandwidth is kept the same,
the capacity of caches should
be quadrupled to achieve
similar performance for each
tread [4]

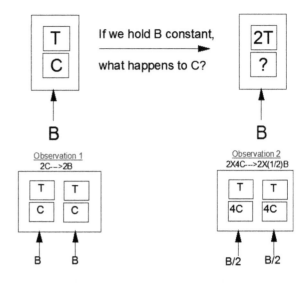

not doubled, then each cache must be quadrupled to make up for the net bandwidth loss per thread [4], as seen in the bottom right of Fig. 1.1.

1.1.2 Increasing Power Consumption

The power dissipation has always been one of most concerned issues. Although the operation voltage scales down with the processing technology, the power consumption, however, keeps increasing. The integration of more devices and increasing length of global connection result in higher dynamic power consumption. More importantly, as the technology scales down to 90 nm and beyond, the leakage power consumption will catch up with, and may even dominate, dynamic power consumption. This makes leakage power reduction an indispensable component in the nano-era low power design.

As dense cache memories constitute a significant portion of the transistor budget of current processors, leakage optimization for cache memories is of particular importance. It has been estimated that leakage energy accounts for 30 % of L1 cache energy and 70 % of L2 cache energy for a 0.13 μ process [5]. There have been several efforts [5–10] spanning from the circuit level to the architectural level at reducing the cache leakage energy. Circuit mechanisms include adaptive substrate biasing, dynamic supply scaling and supply gating. Many of the circuit techniques have been exploited at the architectural level to control leakage at the cache bank and cache line granularity.

1.1.3 Vulnerability to Soft Errors

As the processing technology scales down, the reliability is becoming an increasing challenge in high-performance microprocessor design. Due to the continuously reduced feature size, supply voltage and increased on-chip density, the next generation of microprocessors is projected to be more susceptible to soft error strikes [11–13]. Such reliability problems become more server, when the mainstream of processor design moves toward the regime of CMPs, because of two reasons: (1) packaging multiple cores into the same die exposes more devices to soft error strikes [11]; (2) in order to maintain high-performance, the increasing number of cores requires more on-chip memory, which is normally vulnerable to soft errors.

Traditionally, the on-chip memory is implemented with the SRAM technology, which is vulnerable to soft error strikes. In modern processors, the SRAM caches are protected using redundancy information [14–16]. ECC protection mechanisms incurs large area overhead [12, 13]. For example, in order to achieve Single Error Correction Double Error Detection (SECDED), a 64-bit word requires 8 ECC bits as redundant information, which incurs about 12.5 % of area overhead [13, 15]. In addition, the encoding and decoding of ECC codes also consume extra energy for each read or write operation, respectively. Therefore, it is important to minimize its impact on area, power, and even application performance.

1.2 Background of Emerging Memory Technologies

In this section, we provide a brief review of emerging technologies leveraged in our work. We also show that the advantages of these emerging technologies can help mitigate the challenges introduced in the previous section.

Non-volatile memory is defined as memory that can retain the stored information even when not powered. The well-known examples of non-volatile memory include read-only memory, flash memory, etc. In recent years, various emerging NVM technologies are proposed, such as phase change random access memory (PRAM), magnetic random access memory (MRAM) and resistive random access memory (RRAM). The parameters of these technologies are compared with traditional SRAM and DRAM technologies in Table 1.1. These technologies are considered as competitive candidates for future universal memory because of following advantages:

- **Non-volatility** As shown in the definition, the information stored in the memory cell is retained even without power supply.
- **High density** As shown in Table 1.1, the memory cell factors of NVM technologies are much smaller than that of the SRAM technology, therefore, more memory can be integrated if we replace SRAM with NVM technologies.

Table 1.1 Comparison among different memory technologies

	SRAM	DRAM	NAND flash	PRAM	STTRAM	RRAM
Data retention	N	N	Y	Y	Y	Y
Memory cell factor	50–120	6–10	2–5	6–12	4–20	<1
Read time (ns)	1	30	50	20–50	1–10	<50
Write/erase time (ns)	1	50	106–105	50–120	5–20	<100
Write cycles	10^{16}	10^{16}	10^5	10^9	10^{15}	10^{15}
Power (read)	Low	Low	High	Low	Low	Low
Power (write)	Low	Low	High	High	High	Low
Power (standby)	Leakage	Leakage & refresh	None	None	None	None

Note that STTRAM means the second generation of MRAM technology

- **Zero standby power** Because of the non-volatility, no power is needed when memory cells are standby, therefore, there is no standby power for these NVM technologies.
- **Immunity to soft errors** Because of particular information storage mechanisms of NVM technologies, which will be introduced later in this section, it requires much more energy to change the status of a memory cell. Therefore, the data in memory cells will not be effected by particle radiation strikes.

In this book, we focus on two NVM technologies–MRAM and PRAM technologies. The detailed characters and operation mechanisms of the two NVM technologies are introduced in following subsections.

1.2.1 MRAM Technology

The basic difference between the MRAM and the conventional RAM technologies (such as SRAM/DRAM) is that the information carrier of MRAM is a Magnetic Tunnel Junction (MTJ) instead of electric charges [17, 18]. Each MTJ contains *two ferromagnetic layers* and *one tunnel barrier layer*. Figure 1.2 shows a conceptual illustration of a MTJ. One of the ferromagnetic layer (reference layer) has fixed magnetic direction while the other one (free layer) can change its magnetic direction by an external electromagnetic field or a spin-transfer torque. If the two ferromagnetic layers have different directions, the MTJ resistance is high, indicating a "1" state (the anti-parallel case in Fig. 1.2a); if the two layers have the same direction, the MTJ resistance is low, indicating a "0" state (the parallel case in Fig. 1.2b).

The MRAM technology discussed in this work is called *Spin-Transfer Torque RAM* (STTRAM), which is a new generation of MRAM technologies. STTRAMs change the magnetic direction of the free layer by directly passing a spin-polarized current through the MTJ structure. Comparing to the previous generation of MRAMs that uses external magnetic fields to reverse the MTJ status, STTRAMs has the advantage of scalability, which means the *threshold current* to make the status reversal will decrease as the size of the MTJ becomes smaller.

Fig. 1.2 A conceptual view of MTJ structure. **a** Anti-parallel (high resistance), which indicates "1" state; **b** Parallel (low resistance), which indicates "0" state

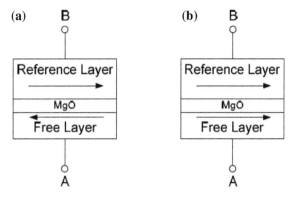

Fig. 1.3 An illustration of an MRAM cell

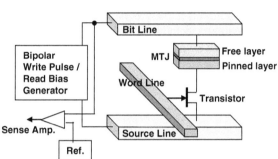

In the STTRAM memory cell design, the most popular structure is composed of one NMOS transistor as the access controller and one MTJ as the storage element ("1T1J" structure) [17]. As illustrated in Fig. 1.3, the storage element, MTJ, is connected in series with the NMOS transistor. The NMOS transistor is controlled by the the word-line (WL) signal. The detailed read and write operations for each MRAM cell is described as follows:

- **Read Operation**: When a *read operation* happens, the NMOS is turned on and a voltage ($V_{BL} - V_{SL}$) is applied between the bit-line (BL) and the source-line (SL). This voltage is negative and is usually very small (-0.1 V as demonstrated in [17]). This voltage difference will cause a current passing through the MTJ, but it is not small enough to invoke a disturbed write operation. The value of the current is determined by the equivalent resistance of MTJs. A sense amplifier compares this current with a reference current and then decides whether a "0" or a "1" is stored in the selected MRAM cell.
- **Write Operation**: When a *write operation* happens, a positive voltage difference is established between SLs and BLs for writing for a "0" or a negative voltage difference is established for writing a "1" . The current amplitude required to ensure a successful status reversal is called threshold current. The current is related to the material of the tunnel barrier layer, the writing pulse duration, and the MTJ geometry [19].

1.2.2 PRAM Technology

Different from the conventional RAM technologies (such as SRAM/DRAM), the information carrier of PRAM is chalcogenide-based materials, such as $Ge_2Sb_2Te_5$ and $Ge_2Sb_2Te_4$ [20]. The crystalline and amorphous states of chalcogenide materials have a wide range of resistivity, about three orders of magnitude, and this forms the basis of data storage. The amorphous, high resistance state is used to represent a bit '0', and the crystalline, low resistance state represents a bit '1'.

Nearly all prototype devices make use of a chalcogenide alloy of germanium, antimony and tellurium (GeSbTe) called GST. When GST is heated to a high temperature (normally over 600 °C), it will get melted and its chalcogenide crystallinity is lost. Once cooled, it is frozen into an amorphous and its electrical resistance becomes high. This process is called RESET. One way to achieve the crystalline state is by applying a lower constant-amplitude current pulse for a time longer than the so-called SET pulse. This is called SET process [21]. The time of phase transition is temperature-dependent. Normally, it takes several 10ns for the RESET and takes about 100ns for the SET [21, 22].

Figure 1.4 gives an illustration of a PRAM cell. The read and write operations for a PRAM cell is described as follows:

- **Read Operation**: To read the data stored in PRAM cells, a small voltage is applied across the GST. Since the SET status and RESET status have a large variance on their equivalent resistance, the data is sensed by measuring the pass-through current. The read voltage is set to be sufficiently strong to invoke detectable current but remains low enough to avoid write disturbance. Like other RAM technologies, each PRAM cell needs an access device for control purpose. As shown in Fig. 1.4, every basic PRAM cell contains one GST and one NMOS access transistor. This structure has a name of "1T1R" where "T" stands for the NMOS transistor and "R"stands for GST. The GST in each PRAM cell is connected to the drain-region of the NMOS in series so that the data stored in PRAM cells can be accessed by wordline controlling.
- **Write Operation**: There are two kinds of PRAM write operations, the SET operation that switches the GST into crystalline phase and the RESET operation that

Fig. 1.4 An illustration of an PRAM cell

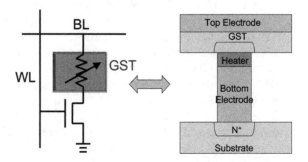

switches the GST into amorphous phase. The SET operation crystallizes GST by heating it above its crystallization temperature, and the RESET operation melt-quenches GST to make the material amorphous [21]. These two operations are controlled by electrical current: high-power pulses for the RESET operation heat the memory cell above the GST melting temperature; moderate power but longer duration pulses for the SET operation heat the cell above the GST crystallization temperature but below the melting temperature. The temperature is controlled by passing through a certain amount of electrical current and generating the required Joule heat.

1.3 Challenges and Prior Related Work

These emerging NVMs also have their own limitations, which impede the adoption of them in the memory hierarchy. These limitations are listed as following:

- **Technology Incompatibility** These emerging memory technologies are incompatible with the traditional CMOS processing technology. Several extra steps are required in the fabrication of these memories. The incompatibility makes it difficult to integrate these NVMs with traditional CMOS design directly, especially for the adoption of these NVMs as on-chip memories.
- **Extra Cost** The extra effort in the process technologies increases the fabrication cost [23, 24]. The extra cost reduces the attraction of using these NVMs as the memory for massive storage (e.g., SSD). Note that, when we integrate these NVMs with traditional CMOS designs on the same chip, the extra steps for NVMs will be applied to the whole die even the memory only takes a part of the design.
- **High Programming Latency and Energy** As shown in Table 1.1, although these NVMs have fast speed for the read operation, the latency and energy of the write operations are much higher than those of SRAM/DRAM. Thus, for those workloads with high intensive write operations, the performance may be degraded and the power consumption may be increased.
- **Write Cycle Issues** As shown in Table 1.1, the write cycle of PRAM is higher than that of NAND flash memory but much lower than those of other memory technologies. Thus, the lifetime issue limits the usage of PRAM, and the wear leveling mechanism may be required.

There has been a lot of research about these NVMs in the device and circuit levels [17–19, 24, 25]. Recently, extensive work has been done in the architectural level.

Desikan et al. first considered on-chip MRAM as the replacement for DRAM memories [26, 27], but the MRAM they have used is based on old generation of MRAM technology, which has scalability issue. Dong et al. proposed to use the second generation of MRAM as the replacement of SRAM caches for single processor [24]. Some other research has also been done to leverage STTRAM as on-chip caches for CMPs [28–30]. Ipek et al. proposed the "resistive computation", which explored STTRAM based on-chip memory and combinational logic in processors to avoid the power wall [31]. Since the STTRAM is immune to radiation based soft errors, the trade-off among reliability, performance, and power consumption is explored by using STTRAM as different levels of on-chip memory [32].

Because of low performance and the endurance problems, it is not feasible to use PRAM as on-chip memory with intensive accesses. Instead, PRAM is proposed to replace DRAM as main memory [33–35], or is used as the lowest level on-chip memory such as L3 caches [29]. Some approaches of hybrid DRAM/PRAM memory have been proposed to leverage the advantages of two memory technologies [36, 37]. In order to improve the lifetime of PRAM memory, some research has been done for write intensity reduction and wear-leveling mechanisms [35, 38–41]. In addition, different error correction mechanisms for hard errors of PRAM have been studied. Besides, PRAM is also used in together with NAND flash memory to improve the performance and lifetime of SSD and reduce the power consumption at the same time [42, 43].

1.4 Goals of this Book

This work intends to leverage the advantages of the emerging memory technologies in the architectural level, in order to achieve high performance, low power, and high reliability memory hierarchy. As we have introduced in Sect. 1.3, it is not straightforward to use these emerging NVMs in the memory hierarchy because of those challenges. The following questions must be considered and answered to achieve the design goals:

- *How to choose the proper memory technologies in the design of different levels of the memory hierarchy?*
- *How to improve the architecture of the traditional memory hierarchy to facilitate the adoption of these emerging memories?*
- *How to leverage the advantages of different memory technologies?*

In this book, we intend to answer these questions by exploring the memory hierarchy design from different angles. First, we propose to replace different levels of a traditional memory hierarchy with proper emerging memory technologies, according to their characters. We will show that both performance and power consumption can be improved after we apply modifications to the memory architectures. At the same time, we find it feasible to leverage advantages from different memory technologies by using the hybrid memory design. Second, we presents a analytical model named "Moguls" to theoretically study the optimization design of a memory hierarchy. With the help of this model, we can not only estimate the level number and the corresponding capacity of the memory hierarchy, but also find out the best choice of memory technology in each level quickly. Third, we explore the vulnerability of the CMPs to radiation-based soft errors by replacing different levels of on-chip memory with STTRAMs. The results show the trade-off among reliability, performance, and power consumption.

1.5 The Organization of this Book

The remainder of the book is organized as follows. Chapter 2 presents our methods of replacing various NVMs in different levels of the memory hierarchy and introduces the modifications of memory architectures to facilitate the adoptions of NVMs. In Chapter 3, we introduce the Moguls model and show how to find an optimized memory hierarchy with either single or multiple memory technologies. The impacts of using STTRAM on vulnerability to radiation based soft errors are studied Chapter 4. Finally, we conclude our work in Chapter 5.

References

1. Albonesi, D.H., Koren, I.: Improving the memory bandwidth of highly-integrated, wide-issue, microprocessor-based systems. In: PACT '97: Proceedings of the 1997 International Conference on Parallel Architectures and Compilation, Techniques, pp. 126–135 (1997)
2. Davis, J.D., Laudon, J., Olukotun, K.: Maximizing CMP throughput with mediocre cores. In: PACT '05: Proceedings of the 14th International Conference on Parallel Architectures and Compilation, Techniques, pp. 51–62 (2005)
3. Burger, D., Goodman, J.R., Kagi, A.: Limited bandwidth to affect processor design. Micro IEEE **17**(6), 55–62 (1997)
4. Emma, P.: The end of scaling? Revolutions in technology and microarchitecture as we pass the 90 nanometer node. SIGARCH Comput. Archit. News **34**(2), 128 (2006). http://doi.acm.org/10.1145/1150019.1136496
5. Powell, M., Yang, S.H., Falsafi, B., Roy, K., Vijaykumar, T.N.: Reducing leakage in a high-performance deep-submicron instruction cache. IEEE Trans. Very Large Scale Integr. Syst. **9**(1), 77–90 (2001). http://dx.doi.org/10.1109/92.920821
6. Kuroda, T., Sakurai, T.: Threshold-voltage control schemes through substrate-bias for low-power high-speed cmos lsi design. J. VLSI Signal Process. Syst. **13**(2–3), 191–201 (1996). http://dx.doi.org/10.1007/BF01130405
7. Zhou, H., Toburen, M.C., Rotenberg, E., Conte, T.M.: Adaptive mode control: a static-power-efficient cache design. In: PACT '01: Proceedings of the 2001 International Conference on Parallel Architectures and Compilation Techniques, p. 61. IEEE Computer Society, Washington, DC (2001)
8. Meng, Y., Sherwood, T., Kastner, R.: Exploring the limits of leakage power reduction in caches. ACM Trans. Archit. Code Optim. **2**(3), 221–246 (2005). http://doi.acm.org/10.1145/1089008.1089009
9. Golubeva, O., Loghi, M., Macii, E., Poncino, M.: Locality-driven architectural cache sub-banking for leakage energy reduction. In: ISLPED '07: Proceedings of the 2007 International Symposium on Low Power Electronics and Design, pp. 274–279. ACM, New York (2007). http://doi.acm.org/10.1145/1283780.1283839
10. Kim, N.S., Flautner, K., Blaauw, D., Mudge, T.: Circuit and microarchitectural techniques for reducing cache leakage power. IEEE Trans. Very Large Scale Integr. Syst. **12**(2), 167–184 (2004). http://dx.doi.org/10.1109/TVLSL.2003.821550
11. Zhang, W., Li, T.: Managing multi-core soft-error reliability through utility-driven cross domain optimization. In: Proceedings of ASAP, pp. 132–137 (2008)
12. Yoon, D.H., Erez, M.: Memory mapped ECC: low-cost error protection for last level caches. In Proceedings of ISCA, pp. 116–127 (2009). http://doi.acm.org/10.1145/1555815.1555771
13. Kim, S.: Reducing area overhead for error-protecting large L2/L3 caches. IEEE Trans. Comput. **58**(3), 300–310 (2009). http://dx.doi.org/10.1109/TC.2008.174

14. Bossen, D., Tendler, J., Reick, K.: Power4 system design for high reliability. IEEE Micro **22**, 16–24 (2002)
15. Quach, N.: High availability and reliability in the itanium processor. IEEE Micro **20**, 61–69 (2000)
16. Phelan, R.: Addressing Soft Errors in Arm Core-Based SoC. ARM, Cambridge (2003)
17. Hosomi, M., Yamagishi, H., Yamamoto, T., Bessho, K., Higo, Y., Yamane, K., Yamada, H., Shoji, M., Hachino, H., Fukumoto, C., Nagao, H., Kano, H.: A novel non-volatile memory with spin torque transfer magnetization switching: Spin-RAM. In: International Electron Devices Meeting, pp. 459–462 (2005)
18. Zhao, W., Belhaire, E., Mistral, Q., Chappert, C., Javerliac, V., Dieny, B., Nicolle, E.: Macro-model of spin-transfer torque based magnetic tunnel junction device for hybrid magnetic-CMOS design. In: IEEE International Behavioral Modeling and Simulation, Workshop, pp. 40–43 (2006)
19. Diao, Z., Li, Z., Wang, S., Ding, Y., Panchula, A., Chen, E., Wang, L.C., Huai, Y.: Spin-transfer torque switching in magnetic tunnel junctions and spin-transfer torque random access memory. J. Phys. Condens. Matter **19**(16), 165, 209 (13 pp) (2007)
20. Kang, D., Ahn, D., Kim, K., Webb, J., Yi, K.: One-dimensional heat conduction model for an electrical phase change random access memory device with an $8f^2$ memory cell (f=0.15 μm). J. Appl. Phys. **94**, 3536–3542 (2003). 10.1063/1.1598272
21. Hudgens, S.: OUM nonvolatile semiconductor memory technology overview. In: Proceedings of Materials Research Society Symposium (2006)
22. Zhang, Y., et al.: An integrated phase change memory cell with Ge nanowire diode for cross-point memory. In: Proceedings of IEEE Symposium on VLSI Technology, pp. 98–99 (2007).10.1109/VLSIT.2007.4339742
23. Lam, C.: Cell design considerations for phase change memory as a universal memory. In: Proceedings of International Symposium on VLSI Technology, Systems and Applications, pp. 132–133 (2008).10.1109/VTSA.2008.4530832
24. Dong, X., Wu, X., Sun, G., Xie, Y., Li, H., Chen, Y.: Circuit and microarchitecture evaluation of 3D stacking magnetic RAM (MRAM) as a universal memory replacement. In: DAC '08: Proceedings of the 45th Annual Conference on Design Automation, pp. 554–559 (2008)
25. Desikan, R., Lefurgy, C.R., Keckler, S.W., Burger, D.: On-chip MRAM as a High-Bandwidth Low-Latency Replacement for DRAM Physical Memories. Technical report (2002)
26. Davis, W.R., Wilson, J., Mick, S., Xu, J., Hua, H., Mineo, C., Sule, A.M., Steer, M., Franzon, P.D.: Demystifying 3D ICs: the pros and cons of going vertical. IEEE Des. Test Comput. **22**(6), 498–510 (2005)
27. Xie, Y., Loh, G.H., Black, B., Bernstein, K.: Design space exploration for 3D architectures. ACM J. Emerg. Technol. Comput. Syst. **2**(2), 65–103 (2006)
28. Sun, G., Dong, X., Xie, Y., Li, J., Chen, Y.: A novel architecture of the 3D stacked MRAM L2 cache for cmps. In: HPCA 2009. IEEE 15th International Symposium on High Performance Computer Architecture, 2009, pp. 239–249 (2009). 10.1109/HPCA.2009.4798259
29. Wu, X., Li, J., Zhang, L., Speight, E., Rajamony, R., Xie, Y.: Hybrid cache architecture with disparate memory technologies. In: Proceedings of the 36th Annual International Symposium on Computer Architecture, pp. 34–45 (2009). http://doi.acm.org/10.1145/1555754.1555761
30. Zhou, P., Zhao, B., Yang, J., Zhang, Y.: Energy reduction for stt-ram using early write termination. In: Proceedings of the 2009 International Conference on Computer-Aided Design, ICCAD '09, pp. 264–268. ACM, New York (2009). http://doi.acm.org/10.1145/1687399.1687448
31. Guo, X., Ipek, E., Soyata, T.: Resistive computation: avoiding the power wall with low-leakage, stt-mram based computing. In: Proceedings of the 37th Annual International Symposium on Computer Architecture, ISCA '10, pp. 371–382. ACM, New York (2010). http://doi.acm.org/10.1145/1815961.1816012
32. Sun, G., Kursun, E., Rivers, J., Xie, Y.: Improving the vulnerability of cmps to soft errors with 3d stacked non-volatile memory. In: Proceedings of ICCD (2011)

33. Zhou, P., Zhao, B., Yang, J., Zhang, Y.: A durable and energy efficient main memory using phase change memory technology. In: Proceedings of the 36th Annual International Symposium on Computer Architecture, pp. 14–23 (2009). http://doi.acm.org/10.1145/1555754.1555759

34. Lee, B.C., Ipek, E., Mutlu, O., Burger, D.: Architecting phase change memory as a scalable dram alternative. In: Proceedings of the 36th Annual International Symposium on Computer Architecture, pp. 2–13 (2009). http://doi.acm.org/10.1145/1555754.1555758

35. Qureshi, M.K., Srinivasan, V., Rivers, J.A.: Scalable high performance main memory system using phase-change memory technology. In: Proceedings of the 36th Annual International Symposium on Computer Architecture, pp. 24–33 (2009). http://doi.acm.org/10.1145/1555754.1555760

36. Park, H., Yoo, S., Lee, S.: Power management of hybrid dram/pram-based main memory. In: Design Automation Conference (DAC), 2011 48th ACM/EDAC/IEEE, pp. 59–64 (2011)

37. Liu, T., Zhao, Y., Xue, C.J., Li, M.: Power-aware variable partitioning for d with hybrid pram and dram main memory. In: Design Automation Conference (DAC), 2011 48th ACM/EDAC/IEEE, pp. 405–410 (2011)

38. Cho, S., Lee, H.: Flip-n-write: a simple deterministic technique to improve pram write performance, energy and endurance. In: Proceedings of MICRO 2009, pp. 347–357. http://doi.acm.org/10.1145/1669112.1669157

39. Qureshi, M.K., et al.: Enhancing lifetime and security of pcm-based main memory with startgap wear leveling. In: Proceedings of MICRO 2009, pp. 14–23. http://doi.acm.org/10.1145/1669112.1669117

40. Seong, N.H., Woo, D.H., Lee, H.H.S.: Security refresh: prevent malicious wear-out and increase durability for phase-change memory with dynamically randomized address mapping. SIGARCH Comput. Archit. News **38**, 383–394 (2010). http://doi.acm.org/10.1145/1816038.1816014

41. Sun, G., Niu, D., Ouyang, J., Xie, Y.: A frequent-value based pram memory architecture. In: Design Automation Conference (ASP-DAC), 2011 16th Asia and South Pacific, pp. 211–216 (2011).10.1109/ASPDAC.2011.5722186

42. Kim, J., Lee, H., Choi, S., Bahng, K.: A PRAM and NAND flash hybrid architecture for high-performance embedded storage subsystems. In: Proceedings of ACM international conference on Embedded software, pp. 31–40 (2008). http://doi.acm.org/10.1145/1450058.1450064

43. Park, Y., Lim, S., Lee, C., Park, K.: PFFS: a scalable flash memory file system for the hybrid architecture of phase-change RAM and NAND flash. In: Proceedings of ACM Symposium on Applied Computing (2008)

Chapter 2
Replacing Different Levels of the Memory Hierarchy with NVMs

2.1 Introduction

As the first step of exploring the usage of various NVMs in different levels of the memory hierarchy, we compare the NVMs with memories used in the traditional memory hierarchy, as shown in Fig. 2.1. Then, we try to replace those SRAM/DRAM/flash based memories with STTRAM/PRAM and propose proper modifications in the architecture level.

First, we evaluate the benefits of stacking STTRAM L2 caches atop CMPs. We develop a cache model for stacking STTRAM and then compare the STTRAM-based L2 cache against its SRAM counterpart with the similar area, in terms of performance and energy. The comparison shows that: (1) For applications that have moderate write intensities to L2 caches, the STTRAM-based cache can reduce the total cache power significantly because of its zero standby leakage and achieve considerable performance improvement because of its relatively higher cache capacity; (2) For applications that have high write intensities to L2 caches, the STTRAM-based cache can cause performance and power degradation due to the long latency and the high energy of STTRAM write operations. These two observations imply that STTRAM-based caches might not work efficiently if we directly introduce them into the traditional CMP architecture because of their disadvantages on write latency and write energy. In light of this concern, we propose two architectural techniques, read-preemptive write buffer and SRAM-STTRAM hybrid L2 cache, to mitigate the STTRAM write-associated issues. The simulation result shows that performance improvement and power reduction can be achieved effectively with our proposed techniques even under the write-intensive workloads.

Second, we present a frequent-value based data storage architecture for PRAM memory. Static and dynamic profiling methods are presented to identify frequent values for different applications. With this architecture, the write intensity to PRAM memory is studied at the "data-level" instead of at the "bit-level". Through exploring the frequent-value locality, the data, which are frequently written back to PRAM memory, are stored with a compressed (encoded) form. Consequently, the write

G. Sun, *Exploring Memory Hierarchy Design with Emerging Memory Technologies*,
Lecture Notes in Electrical Engineering 267, DOI: 10.1007/978-3-319-00681-9_2,
© Springer International Publishing Switzerland 2014

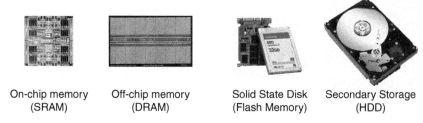

| On-chip memory | Off-chip memory | Solid State Disk | Secondary Storage |
| (SRAM) | (DRAM) | (Flash Memory) | (HDD) |

Fig. 2.1 An illustration of the traditional memory hierarchy

intensity to PRAM memory is significantly reduced. In addition, since such approach is achieved at the data-level, the frequent-value based storage can be used in parallel with those bit-level methods seamlessly to further improve the endurance of PRAM memory.

Third, we employ PRAM as the log region of our hybrid storage. We propose to use PRAM as the log region of the NAND flash memory storage system and develop a set of management policies for PRAM log region to fully exploit its advantages of in-place updating ability, fine access granularity, long endurance, etc. The hybrid architecture has the following advantages: (1) the ability of in-place updating can significantly improve the usage efficiency of log region by efficiently eliminating the out-of-date log data; (2) the fine-granularity access of PRAM can greatly reduce the read traffic from the NAND flash memory to the DRAM data buffer since the size of logs loaded for the read operation is decreased; (3) the energy consumption of the storage system is reduced as the overhead of writing and reading log data is decreased with the PRAM log region; (4) the lifetime of the NAND flash memory could be increased because the number of erase operations are reduced, and the endurance of the PRAM log region can be traded off to further improve the performance of the storage system. The simulation results show that, with proper PRAM management policies, both performance and endurance of the hybrid storage are significantly improved.

2.2 3D Stacked STTRAM L2 Caches

STTRAM is a promising memory technology, which has fast read access, high density, and non-volatility. Using 3D heterogeneous integration, it becomes feasible and cost-efficient to stack STTRAM atop conventional chip multiprocessors (CMPs). However, one disadvantage of STTRAM is its long write latency and its high write energy. In this section, we first stack STTRAM-based L2 caches directly atop CMPs and compare it against SRAM counterparts in terms of performance and energy. We observe that the direct STTRAM stacking might harm the chip performance due to the aforementioned long write latency and high write energy. To solve this problem, we then propose two architectural techniques: read-preemptive write buffer and SRAM-STTRAM hybrid L2 cache. The simulation result shows that our optimized STTRAM L2 cache improves performance by 4.91 % and reduces power by 73.5 % compared to the conventional SRAM L2 cache with the similar area.

2.2.1 Modeling an STTRAM Based Cache

To model the STTRAM-based cache, the first step is to estimate the area of a STTRAM cell. As shown in Fig. 1.3, each STTRAM cell is composed of one NMOS transistor and one MTJ. Note that the new generation of MRAM (STTRAM) technology offers scalability for MTJ, the size of MTJ is only limited by manufacture techniques. However, the NMOS transistor has to be sized properly so that it can pass sufficient current for the MTJ to change the cell status. Since the current driving ability of NMOS transistor is proportional to its W/L ratio, the W/L ratio becomes the key point to determine the STTRAM cell size.

By using HSPICE simulation, we find that the minimum W/L ratio for the NMOS transistor in 65 nm technology should be 10 in order to drive the required threshold current of 195 μA for the status reversal. We assume the width of the source or drain regions of an NMOS transistor is 1.5 F, where F is the feature size. Combining these requirements together, the size of the minimum STTRAM cell we can design is set to $10 F \times 4 F$, which equals to $40 F^2$. The specification of our targeted STTRAM cell is listed in Table 2.1.

Despite the physical mechanism in the storage cell, STTRAM and SRAM caches have almost the same peripheral interfaces from the circuit designer's point of view. In 65 nm technology, the cell area simulation results show that the area of one STTRAM cell is about 25 % of one SRAM cell ($146 F^2$ extracted from CACTI). Table 2.2 shows the comparison of area, access time and access energy for a 512 KB STTRAM cache bank and a 128 KB SRAM bank, which are used in this work.

Table 2.1 STTRAM cell specifications

Technology	Write pulse duration	Threshold current	Cell size	Aspect ratio
65 nm	10 ns	195 μA	$40 F^2$	2.5

Table 2.2 Area, access time and energy comparison (65 nm technology)

Cache size	Area (mm^2)	Read latency (ns)	Write latency (ns)	Read energy (nJ)	Write energy (nJ)
128 KB SRAM	3.62	2.252	2.264	0.895	0.797
512 KB STTRAM	3.30	2.318	11.024	0.858	4.997

2.2.2 Configurations and Assumptions

As the caches size increases, the wire delay in deep sub-micron regions has made the Non-Uniform Cache Architecture (NUCA) [1] more attractive than the one with uniform access latency. In NUCA, the cache is divided into multiple banks with different access lattices according to their locations relative to cores. These individual banks can be connected through a mesh-based interconnection network called Network-on-Chip (NoC).

Based on the latest 2D cache simulator CACTI [2], we developed our NoC-based 3D NUCA cache model, which can incorporate the features of 3D integration. The key concept is to use NoC routers for communications within a 2D layer, while using a special through-silicon-bus (TSB) to communicate in the vertical direction between layers. We also elaborated low-level parameters such as the latency and the energy of interconnection wires and modified the technology parameters to be consistent with the STTRAM technology.

Figure 2.2a illustrates an example of the 3D NUCA structure. There are four cores located in the layer 1. In each core, there is a cache controller connected to a through-silicon-bus (TSB) from which data are moved through layers between processing cores and caches. The TSBs are implemented with through-silicon-vias (TSVs). This bus structure has the advantage of short connections provided by 3D integration. It has been reported that the vertical latency for traversing the height of a 20-layer stack is only 12 ps [3]. The latency of traversing on a TSB through one layer is negligible compared to the one traversing between two NoC routers in 2D. Consequently, it is feasible to have single-hop communication with these buses because of the short distance between the layers. Using the vertical buses, the processing cores can have the same access latency to banks in the same location of

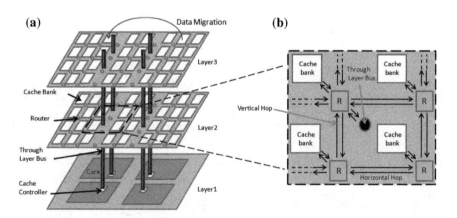

Fig. 2.2 **a** An illustration of the proposed 3D NUCA structure, which includes 1 core layer, 2 cache layers. There are 4 processing cores per core layer, 32 cache banks per cache layer, and 4 through-layer-bus across layers; **b** Connections amongst routers, caches banks and through-layer-buses

different layers. In addition, hybridization of the NoC router with the bus requires only one additional link (instead of two) on the NoC router, because the bus is a single entity for communicating both up and down [4].

As shown in Fig. 2.2a, L2 caches are placed on top of cores and they are located on layer 2 and layer 3. There are 32 cache banks in each layer which are connected with network on-chip routers. Each cache bank can be implemented with either SRAMs or STTRAMs. Figure 2.2b shows a detailed 2D structure for a cache layer. Three out of the four routers have five input/output ports connecting the four neighbor routers and the internal cache bank, while the fourth router needs an extra input/output port to communicate with the through-silicon-bus (TSB).

Similar to prior approaches [4–6], the proposed model supports the migration of moving data closer to the accessing core. For set associative cache, the cache ways belonging to the same index should be distributed into different banks in order to implement the data migration. In our model, each cache layer is equally divided into several zones. The number of zones is equal to the number of cores, and each zone has a TSB located at its center. The cache ways of each set are uniformly distributed into these zone. This architecture promises that, within each cache set, there are several ways of cache lines close to the active core. Figure 2.3 gives an illustration of distributing eight ways into four zones. Note that, in this architecture, one cache request should be sent to all zones so that all caches ways are accessed. It will increase the network congestion which are also modeled as a part of router latency. The migration policy and cache banks placements are tailored to the 3D architecture so that data migrations are limited to the same layer. This limitation help reduce the workloads of TSBs. Figure 2.2a shows an example of data migration after which the core in the upper-left corner can access the data faster. Note that this data migration policy is called *inter-migration* in this work, in order to differentiate another migration policy introduced later.

The *memory(caches) + logic(cores)* structure is employed in our model as shown in Fig. 2.2a. The advantages of this structure are: (1) placing L2 caches in separate layers makes it possible to integrate STTRAM with traditional CMOS process technology, (2) separating cores from caches simplifies the design of the TSB and routers because the TSB is now connected to cache controllers directly, and there is no direct connection between routers and cache controllers.

We provide one through-layer-bus for each core in the model. This aggressive structure gains more benefits from the high band-width advantages of 3D stacking. It is reported that TSVs have pitches of only $4 - 10\,\mu m$ [3]. Even at the high-end with

Fig. 2.3 Eight caches ways are distributed in four banks. Assume four cores and accordingly four zones each layer

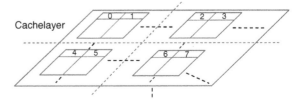

a 10 μm TSV-pitch, a 1024-bit bus (much wider than our TSB) would only require an area of 0.32 mm². In our model, the die area of an 8-core CMP is assumed to be 60 mm² (introduced later). Therefore, it is feasible to assign one TSB for each core.

Our baseline configuration is an 8-core in-order processor using the Ultra Sparc-III ISA. We estimate the area of SRAM cache banks with CACTI6.0 [2]. For STTRAM cache banks, we extract low-level parameters of wire connections and routers from CACTI, and input them to our STTRAM cell model to estimate the area. Currently, there is no tool to predict the area of processing cores. We investigate real hardware, such as Cell Processor, Sun UltraSPARC T1, etc. [7, 8], and estimate the area of eight processing cores to be 60 mm². Furthermore, we assume that one cache layer fits to either a 2 MB SRAM or an 8 MB STTRAM L2 cache. The configurations are detailed in Table 2.3. Note that the power of each core is also estimated based on data sheets from real products [7, 8].

We use the Simics toolset [9] for performance simulations. Our 3D NUCA architecture is implemented as an extended module in Simics. We use a few multi-threaded benchmarks from the *OpenMP2001* [10] and *PARSEC* [11] suites, as shown in Table 2.4. These benchmarks are chosen so that we have a range of transaction intensities to the L2 caches, since the performance and power of STTRAM caches are closely related to transaction intensity. The average numbers of total transactions and write transactions to L2 caches per 1K instructions are listed in Table 2.4. We pin

Table 2.3 Baseline configuration parameters

Processors	
Number of cores	8
Frequency	3 GHz
Power	6 W/core
Issue width	1 (in order)
Memory parameters	
L1 (private I/D)	16 + 16 KB, 2-way, 64B line, 2-cycle, write-through, 1 read/write port
SRAM L2 (shared)	2 MB (16 × 128 KB), 32-way, 64 B line, read/write per bank : 7-cycle, write-back, 1 read/write port
STTRAM L2 (shared)	8 MB (16 × 512 KB), 32-way, 64 B line, read penalty per bank: 7-cycle, write penalty per bank: 33-cycle, write-back, 1 read/write port
Write buffer	4 entry, retire-at-2
Main memory	4 GB, 500-cycle latency
Network parameters	
Number of layers	2
Number of TSB	8
Hop latency	TSB: 1 cycle V_hop: 1 cycle; H_hop: 1 cycle
Router latency	2-cycle

Table 2.4 L2 transaction intensities

Name	galgel	apsi	equake	fma3d	swim	streamcluster
TPKI	1.01	4.15	7.94	8.43	19.29	55.12
WPKI	0.31	1.85	3.84	4.00	9.76	23.326

TPKI is the number of total transactions per 1K instructions and WPKI is the number of write transactions per 1K instructions

one thread on one core during the simulation. For each simulation, we fast forward to warm up the caches, and then run three billion cycles in detailed mode.

2.2.3 Replacing SRAM with STTRAM as L2 Caches

In this section, we replace of a SRAM L2 cache with a STTRAM L2 cache that has the similar area. In this strategy, we integrate as many caches in the cache layers as possible. In a 2M SRAM cache layer that has 16×128 KB banks, the SRAM banks can be replaced by a STTRAM cache with 16×512 KB banks, with the similar average on-chip-network access latency for each bank (as shown in Table 2.2).

2.2.3.1 Performances Analysis

Because the number of banks remains the same, the average latency of read operations to the SRAM cache and the STTRAM cache are similar. Since the capacity of the STTRAM L2 cache is much larger, the access miss rate to the L2 cache should decrease. The normalized access miss rates of L2 caches (a comparison between using the 2M SRAM L2 cache and the 8M STTRAM L2 cache) are listed in Fig. 2.4. On average, the miss rates are reduced by 19.0 and 12.5 % for SNUCA STTRAM cache and DNUCA STTRAM cache, respectively. Note that the miss rates of using SNUCA and DNUCA caches are compared separately. It is because the data migration in our DNUCA model evicts the data in the destination cache line, and the miss

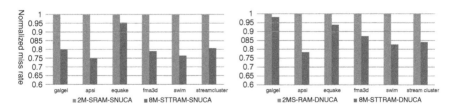

Fig. 2.4 The comparison of L2 caches access miss rates for SRAM L2 cache and STTRAM L2 cache that have similar area. Larger capacity of STTRAM cache results in smaller cache miss rates

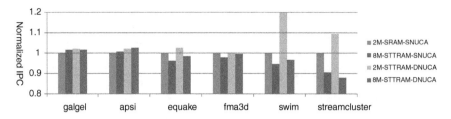

Fig. 2.5 IPC using different configurations of STTRAM and SRAM caches (Normalized by that of SRAM)

rate increases. However, the benefits of moving data close to processing cores will improve the performance. Thus, we compare the miss rates of using SRAM and STTRAM caches in different architectures separately.

The IPC numbers of running different benchmarks are listed in Fig. 2.5. Although the miss rates L2 cache decrease when we use the STTRAM cache, resulting a small performance improvement for the first two benchmarks ("galgel" and "apsi"), but the performance for the last four benchmarks is not improved as expected. On average, the performance degradation are about 3.09 and 7.52 % for SNUCA STTRAM and DNUCA STTRAM, respectively. From Table 2.4, one can observe that the intensities of write operations (WPKI) of the first two benchmarks are much lower than that of the last four benchmarks. It means that a smaller L2 miss rate of using a large STTRAM cache can bring benefits for certain, especially when the intensity of write operations is low. However, when the intensity of write operations is high, the performance penalty due to long latency associated with write operations offsets the benefits from smaller miss rates due to the larger capacity of STTRAM cache. This observation is further supported by comparing the results of using SNUCA and DNUCA. From Fig. 2.5, one can observe that performance degradation is more significant when we use DNUCA STTRAM cache. It is because data migrations in DNUCA incur more write operations, which cause more penalties when using the STTRAM cache.

To summarize, we conclude our first observation of using the STTRAM cache as:

Observation 1 *Replacing the SRAM L2 cache with a STTRAM cache, which has the similar area but with a large capacity, can reduce the access miss rate of the L2 cache. However, the long latency associated with the write operations to the STTRAM cache has a negative impact on the performance. When there is a high intensity of write operations to the STTRAM cache, the benefits of smaller miss rate could be totally eliminated by the penalties due to write operations, eventually resulting in performance degradation.*

2.2.3.2 Power Analysis

The major contributors of the total power consumption in caches are leakage power and dynamic power:

- *Leakage Power:* When process technology scales down to sub-90 μm, the leakage power in CMOS technology becomes dominant. Since STTRAM caches use non-volatile MJT to store data, it is feasible to turn off the power supply to the STTRAM cells that are not being accessed for leakage power reduction. Figure 2.6 lists the total leakage power of SRAM and STTRAM caches used in this work.
- *Dynamic Power:* The calculation of the dynamic power for the NUCA cache is described as follows. For each transaction, the total dynamic power is composed of the memory cell access power, the router access power, and the power consumed by wire connections. In this work, these values are either simulated by HSPICE or extracted from the CACTI [2]. The access number of routers and the length of wire connections vary based on the location of the requesting core and the requested cache lines. Moreover, the write and the read operation also consumes different amount of dynamic power. Therefore, we need to know the exact number of read and write transactions in order to get the average dynamic power.

Figure 2.7 shows the comparison of the total power for SRAM and STTRAM L2 caches. One can observe that:

- For all benchmarks, the total power of the STTRAM caches is smaller than that of the SRAM cache. The average power savings across all benchmarks are about 78 and 68 % for SNUCA and DNUCA, respectively. The power saving for DNUCA STTRAM is smaller, because more write operations due to data migration cause extra dynamic power consumption. It is obvious that the "low leakage power" feature makes the STTRAM more attractive to be used as large on-chip memory, especially when leakage power dominates STTRAM total power consumption when technology scales.

Fig. 2.6 Leakage power of SRAM and STTRAM caches at 80 °C

Cache Configuration	Leakage Power
2M SRAM Cache	2.089W
2M STTRAM Cache	0.255W

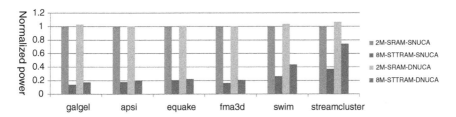

Fig. 2.7 Total power of the SRAM caches and STTRAM cache (Normalized by the power of 2 MB SNUCA SRAM cache)

- The average power saving for the first four benchmarks is more than 80%, due to low intensity of write operations to L2 cache. However, for the last benchmark (*"streamcluster"*), the total power savings are about 63 and 30% for SNUCA and DNUCA, respectively. The saving is much smaller than the average across all benchmarks, because it has much higher L2 cache write intensity than other benchmarks (see Table 2.4). The high energy associated with the write operations in STTRAM makes the total power savings not as significant as other benchmarks.
- For SRAM L2 cache, since leakage power dominates, the total power for SNUCA SRAM and DNUCA SRAM are very close. On the contrary, the dynamic power may dominate in the STTRAM cache because of its low leakage power and high energy of the write operation. Consequently, for benchmarks with high intensity of write operations to L2 (such as *"swim"* and *"streamcluster"*), the total power of DNUCA STTRAM cache could be much higher than that of SNUCA STTRAM cache.

Note that thermal issues are one of the major concerns for the adoption of 3D integration technology. A large power saving introduced by replacing SRAM L2 cache with STTRAM L2 cache implies that it is good for thermal management in 3D stacking chip, especially when multiple layers of memory are stacked on CMPs.

To summarize, our second conclusion of using the STTRAM cache is:

Observation 2 *Replacing the SRAM L2 cache with a STTRAM cache, which has similar area but with larger capacity, can greatly reduce the leakage power, resulting a significant total power saving when leakage power dominates. However, when there are intensive write operations, the dynamic power increase significantly because of the high energy associated with write operations to the STTRAM cache, and the total power saving could be reduced.*

The two conclusions show that, if we directly replace the SRAM cache with the STTRAM cache in a straightforward way, the performance and power penalties caused by the long latency and high energy associated with the write operations in STTRAM cache can offset the performance and power benefits of using the STTRAM cache when there are high intensity of write operations to the L2 cache.

2.2.4 Novel 3D-Stacked Cache Architecture

In this section we propose two techniques in order to mitigate the write operation problem of using STTRAM caches. In 2.2.4.1, an preemptive write buffer is employed to reduce the stall time caused by the long latency of the write operation to the STTRAM cache. In 2.2.4.2, an approach of SRAM-STTRAM hybrid L2 cache is proposed to reduce the number of write operations to the STTRAM cache so that we could improve both the performance and the power. We combine these two techniques together as on optimized architecture for the STTRAM cache.

2.2.4.1 A Read-preemptive Write Buffer

The first conclusion shows that the long latency of a write operation to the STTRAM has a serious impact on the performance. In the scenario that a write operation is followed by several read operations, the write operation may block the read operations, and the performance degrades. If these read operations do not access the same cache line as the write operation does, the read requests can be bypassed to the cache. In the modern processor, a write buffer is usually employed to bypass the read request. As listed in Table 2.3, when we use the SRAM cache, a write buffer with four entries works well enough. However, the experimental results in Sect. 2.2.3.1 show that this write buffer is not fit for the STTRAM cache. In order to make the STTRAM cache work efficiently, we explore the proper size of the write buffer, and propose a read-preemptive management policy for it.

The Structure of the Write Buffer

The structure of the write buffer is shown in Fig. 2.8, and it is operated as follows,

- When there is a write operation, the request is sent into the buffer. If there is a *hit* in the buffer (a valid entry of the same cache line is found in the buffer), the data in the entry are replaced, otherwise, the buffer locates an empty entry to the operation. Note that one entry contains one cache line.
- When there is a read operation, the request is sent to both the buffer and the cache. The buffer searches all valid entries. If there is a read *hit* in the buffer, the buffered data are returned. Otherwise data in the STTRAM cache are returned.
- The buffer executes (retires) the buffered write operations to the STTRAM cache following the retirement policy.

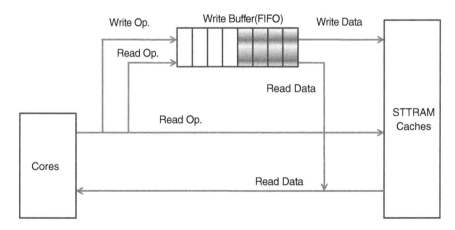

Fig. 2.8 The process structure of cache accesses after adding a write operation buffer

- If the buffer is full, requests from the upper level memory (L1 cache) are blocked till one entry of the buffer is retired.

The Exploration of the Buffer Size

The choice of the buffer size is important. The larger the buffer size, the more write operations could be buffered so that the stall cycles caused by full buffer decrease. However, the more buffered data, the longer time it takes the controller to search them all. The design complexity and the area of the buffer also increase with the buffer size.

Figure 2.9 shows the IPCs improvements of using different size of buffers for two benchmarks. Note that the IPCs are normalized by that of using a STTRAM cache without the write buffer (buffer size equals to 0 entry). Experimental results show that the size of 20 entries is an optimal choice because, for all benchmarks, the performances increase little with the buffer size larger than 20 entries. Compared to the write buffer for the SRAM cache which has four entries, the write buffer for the STTRAM cache requires a much larger size. Note that write buffers in STTRAM caches all have the size of 20 entries in the rest of the section.

A Read-preemptive Policy

Since the L2 cache can receive requests both from the upper level memory (L1 cache) and the write buffer, a priority policy is necessary to solve the conflict that a read request and a write request compete for the right of the execution. For a STTRAM cache, the latency of the write operation is much larger than that of the read one, and the object is to prevent the write operation from blocking the read one. Thus we have:

Rule1: *The read operation always has the higher priority in a competition for the right of the execution.*

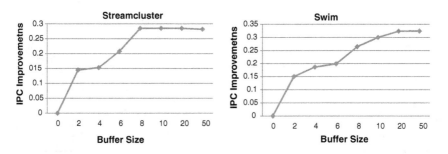

Fig. 2.9 The impact of buffer size. The IPC improvements is normalized by that of using the 2 MB SRAM cache

A read request can also be blocked by a write request that is already in the process of the retirement. The write latency to the STTRAM is so large that a retirement of the write request may block one or more read request for a long period, and the performance degrades. In order to mitigate this problem, we propose a read-preemptive rule as follows:

Rule2: *When a read request is blocked by a write retirement, and the write buffer is not full, the read request can trap and stall the retirement if the preemption condition is satisfied. Then, the read operation obtains the right of the execution to the cache. The stalled retirement will retry later.*

This read-preemptive policy tries to execute the read requests in the STTRAM cache as early as possible. The drawback is that some retirements need to be re-executed, and the possibility of full buffer increases. The pivot is to find a proper preemption condition. One special method is that the retirement is always stalled when a read request is blocked. It means that there is no preemption condition and the read requests are executed to the cache immediately.

Theoretically, if the size of the write buffer is large enough, the read request will never be blocked. However, since the size of the buffer is limited, the increase of the full buffer possibility could also harm the performance. Especially, when the write intensity is high, the non-conditional preemption policy may degrade the performance.

In some cases, it is not a good choice for the read request to stall a write retirement. For example, if the retirement is almost finished, the read request should not stall the retirement process. Consequently, we propose to use the accomplishment degree of the retirement as the preemption condition. Let α denote the accomplishment degree of the retirement, after which the read request will not stall the retirement process. The Fig. 2.10 compares the IPC numbers of using different α in the policy. Note that "$\alpha = 100\,\%$" represents the non-conditional preemption policy and "$\alpha = 0\,\%$" represents the traditional policy. We can find that, for the benchmarks (*galgel* and *apsi*) with low write intensities, the performance increase as the α increases, and the non-conditional preemption policy works best. However, for the benchmark with high

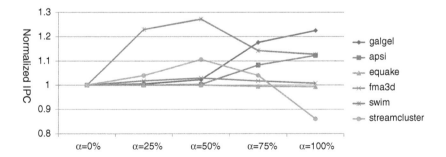

Fig. 2.10 The impact of α on the performances. The IPC numbers are normalized by that of using the traditional policy

intensities of write operations, the performances increase at first, and then decrease as the α increases. Especially, for the benchmark *streamcluster* with the highest write intensity, the performance degrades much with the non-conditional preemption policy. In this work, we set $\alpha = 50\,\%$, and performances of all benchmarks increase with the read-preemptive policy compared to those with the traditional policy.

A counter is requested in order to make the accomplishment degree aware to the cache controller. The counter resets to zero and begins to count the number of cycles when a retirement begins. The cache controller check the counter and decides whether to stall the retirement for the read request. Note that the exact cycles of the write operation could be calculated (except for the case of miss) because the location of the cache bank being accessed is recorded in the write request. However, in order to simplify the design, the cache controller use the average access cycle number instead of calculating the exact number for each request.

The area of the 20 buffer entries can be evaluated as that of a cache whose size is 20×64 Byte (less than 2 KB). We use a 7-bit counter to estimate the accomplishment of the retirement. Since the layer area is fit for a 2 MB SRAM cache. The area overhead is less than 1 %, which is negligible. Similarly, the increase of the leakage power caused by the buffer is also negligible.

Figures 2.11 and 2.12 compares the performances and the power after using our read-preemptive write buffer to those of using the traditional architecture. Compared to the IPC numbers of using the SRAM cache in the baseline configuration, the average performance improvements are 9.93 and 0.41 % for SNUCA and DNUCA respectively. The average power reductions are 67.26 and 59.3 % for SNUCA and DNUCA respectively. Compared to the results in Figs. 2.5 and 2.7, the performances improve a lot, but the power reductions decrease. It is because using our read-preemptive write buffer causes re-executions of some write operations so that the dynamic power increases, and the buffer with larger size also incurs more power.

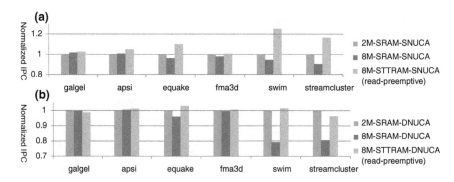

Fig. 2.11 The comparison of IPC among 2M STTRAM, 8M STTRAM with traditional write buffer, and 8M STTRAM with read-preemptive write buffer (Normalized by that of SRAM)

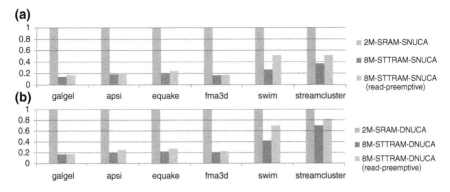

Fig. 2.12 The comparison of total power of 2M SRAM, 8M STTRAM with traditional write buffer, and 8M STTRAM with read-preemptive write buffer (Normalized by that of SRAM)

2.2.4.2 SRAM-STTRAM Hybrid L2 Caches

While the read-preemptive write buffer, which is introduced above, reduces the number of non-critical blocks and hence hides the long latency of STTRAM write operations, the total number of write operations remains the same. In order to reduce the number of write operations to STTRAM cells, in this section, we propose another technique called *SRAM-STTRAM Hybrid Cache* and show how this novel technique can help to mitigate the STTRAM write intensity and further reduce the dynamic power as well as improve the performance.

The Implementation of the Hybrid Cache

The proposed implementation of the hybrid cache is that, instead of building a pure STTRAM cache, we compose the ways in each cache set with a large portion of STTRAM cache lines and a small portion of SRAM cache lines. The main purpose is to keep as many write-intensive data in the SRAM part of caches as possible and thus reduce the number of write operations to the STTRAM part of caches. In this work, for the baseline 32-way set-associative L2 cache design (see Table 2.3), we design an SRAM-STTRAM hybrid L2 cache with 31 ways of STTRAM and 1 way of STTRAM (*31M1S*).

After having these hybrid cache lines, the second step is how to distribute STTRAM cache lines and SRAM ones into separate cache banks. Considering the SRAM part is the minority in the proposed *31M1S* cache, one partitioning alternative is to distribute these SRAM cache lines into different banks so that there are at least several SRAM cache lines close to each processing core. However, this method requires each cache bank as a heterogeneous memory array with SRAM and STTRAM cells and increases the complexity of the cache design. In addition, this distributed partitioning of SRAM cells implies that the SRAM and STTRAM cells have

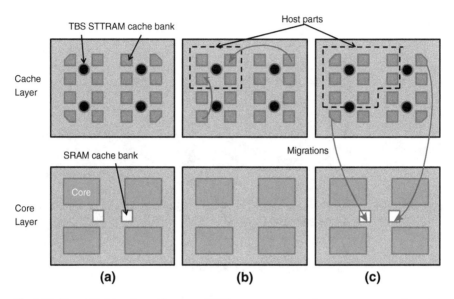

Fig. 2.13 The cache layer from a four-core (TSB) processor used to demonstrate: **a** one placement method of SRAM and STTRAM cache banks, **b** data migrations in original STTRAM caches, **c** data migrations in hybrid STTRAM-SRAM caches

to be fabricated on one layer. Considering the specialization of the STTRAM fabrication process, this method also eliminates the cost advantages of stacking STTRAMs on top of processing cores.

Therefore, we use another alternative that, we reduce the number of cache lines in some STTRAM cache banks compared to the pure STTRAM cache structure (as shown in Fig. 2.13a that the STTRAM banks at four corners are smaller than other STTRAM banks), compensate this cache line loss with SRAM ones, and collect all the SRAM cache lines together to build several entire SRAM banks on the core layer. As shown in Fig. 2.13a, SRAM cache banks are placed in the center of the core layer instead of being distributed. In this method, SRAM and STTRAM cache banks have no difference from the architectural view. The central location of these SRAM cache banks provides moderate access latency from all cores. The implementing of the SRAM-STTRAM hybrid cache using this method only requires the re-design of the NoC topology and does not have any negative impact from the architectural viewpoint. Note that after placing one way of SRAM cache lines in the core layer, the area of the core layer will increase and the area of the cache layer will decrease. In this work, the total size of all the SRAM cache lines is 256 KB, the derived area overhead is about 12.5 %.

The Management Policy of the Hybrid Cache

Another important issue is how to manage the hybrid L2 cache to improve the performance and reduce the power. Because the key point is to reduce the number of write operations to STTRAM cache cells, we need to move as many write-intensive data in SRAM cache banks as possible. The management policy of the hybrid cache can be described as follows:

- The cache controller is aware of the locations of SRAM cache ways and STTRAM cache ways. When there is a write miss, the cache controller first try to place the data in the SRAM cache ways.
- Considering the high probability that a core write data to a specific group of cache lines repeatedly, data in STTRAM caches should be migrated to SRAM caches if the some cache lines are frequently written to. In this work, data in STTRAM caches will be migrated to SRAM caches when they are accessed by two successive write operations. This kind of data migration is named *intra-migration* in order to differentiate *inter-migration* policy introduced in Sect. 2.4.1. Due to the existence of this intra-migration policy, the number of write accesses from cores to STTRAM caches can be reduced.
- Note that read operations from cores are also possible to cause data migrations, the number of which could be even larger than that of direct write accesses from cores. Therefore, a new type inter-migration policy is introduced. Figure 2.13b and c compare the banks from which data can be migrated toward the core in upper-left corner. Figure 2.13b shows that, in original inter-migration policy, the cache layer is divided into 4 uniform groups and there is only one core associative with each part. In this work, banks in each group are named as the *host banks* of their corresponding core. Data can only be migrated from *non-host banks*. For the traditional management policy, the data will be migrated to *host bank*. For the management policy proposed for the hybrid cache, the data can only be migrated to SRAM banks.
  Two data migrations are illustrated in Fig. 2.13b for the traditional inter-migration. When using the hybrid STTRAM-SRAM caches, the *host banks* for a core is redefined as shown in Fig. 2.13c. Two corresponding data migrations are also shown in Fig. 2.13c. Using this policy, there is no data migration between two STTRAM cache lines, which reduces the number of write operations greatly. The drawback is that these SRAM banks are shared for all cores so that their limited sizes may increase L2 miss rates. Note that we have 8M of total cache size, which is considerably large for most applications, our simulation results show that the increase of L2 miss rates is very small.

Figure 2.14 shows the number of write operations to the STTRAM cache per 1K instructions are reduced dramatically by using our hybrid STTRAM-SRAM approach. As a result, the dynamic power associated with write operations to STTRAM cells is reduced, and the performance is improved. Figure 2.15 shows the performance comparison among the proposed STTRAM-SRAM hybrid cache structure and its pure SRAM or pure STTRAM counterparts. Note that the read-

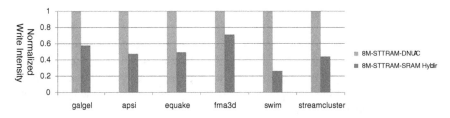

Fig. 2.14 The write intensity to STTRAM of using hybrid STTRAM-SRAM caches compared to that of pure STTRAM caches

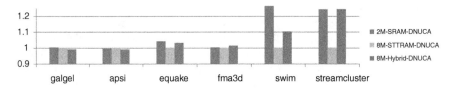

Fig. 2.15 The comparison of IPC among 2M SRAM cache, 8M STTRAM pure cache, and 8M STTRAM-SRAM hybrid cache (Normalized by the IPC of 8M STTRAM pure cache)

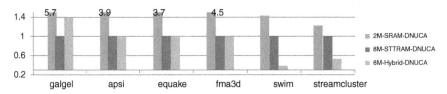

Fig. 2.16 The comparison of total power consumption among 2M SRAM cache, 8M STTRAM pure cache, and 8M STTRAM-SRAM hybrid cache (Normalized by the total power consumption of 8M STTRAM pure cache)

preemptive write buffer is not used here because we want to highlight the benefits of using the hybrid cache structure alone. On average, the hybrid cache structure improves the performance by 5.65 %, which means it mitigates the performance loss of STTRAM caches from 8.48 to 2.61 % compared to their SRAM counterparts.

In Fig. 2.16, we show the comparison in terms of power consumption. Only focusing on the total power consumption difference between the pure STTRAM cache and the SRAM-STTRAM hybrid cache, we find that the total power reduces except for *galgel*. It is because that both the read and the write intensities in *galgel* are so small that the consumed dynamic power is very low. Consequently, the introduction of SRAM cache lines in the hybrid cache brings the leakage power back and this amount of leakage power overhead eliminates the dynamic power reduction achieved by the hybrid structure. However, as the write intensity increases, the STTRAM-SRAM hybrid cache can lower the total power consumption by reducing the dynamic power while paying a small amount of leakage power overhead. For example, the total power consumption is cut by more than half for the applications such as *swim* and *streamcluster*. On average, after the transition from SRAM caches

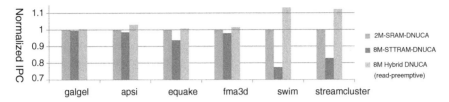

Fig. 2.17 Normalized IPC numbers of using the SRAM cache and the hybrid cache

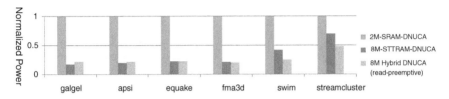

Fig. 2.18 Normalized power of using the SRAM cache and the hybrid cache

to STTRAM ones, this newly-proposed hybrid cache further reduces the total power by 12.45 %.

2.2.4.3 The Combination of Two Techniques

We combine the two techniques together as an optimized architecture for the STTRAM cache. In this architecture, we get more benefits from the advantages of the STTRAM cache, and mitigate the penalties caused by write operations. The performance and power comparisons are listed in Figs. 2.17 and 2.18, respectively. The average IPC number improves by 14.0 and 4.91 % when we compare that of using the STTRAM cache in the optimized architecture to that of using the DNUCA STTRAM and DNUCA SRAM in a traditional architecture respectively. The power consumption reductions are 5.8 and 73.5 % respectively.

2.2.4.4 Temperature Analysis

In the baseline processor of this work, there are only two layers, and the largest cache sizes are 8 MB. However, with the development of CMPs, more L2 caches could be integrated to satisfy the requirement of future applications. Therefore, the heat density will be high enough to be a severe problem in the 3D stacking technology. Consequently, when we decide the maximum limitation of 3D memory stacking, no matter whether it is for SRAM, DRAM, or STTRAM, it is important to estimate the resulting temperature.

In this section, we evaluate the temperature of the 3D memory stacking CMP chip. In the evaluation, the layer of processor cores are allocated right next to heat sink to help heat dissipation. Memory layers (assuming the footprint is equivalent to the area of a 2 MB SRAM) are stacked on top of processor cores and further away from the heat sink. To run this temperature analysis, we choose *ft* from NPB3.2 as the testing benchmark, and use Hotspot 4.1 [12] to evaluate the temperature of the processor.

For L2 caches, the leakage power can be represented with the value in Fig. 2.6. It means all the banks in the cache are powered on. The dynamic power density of caches may vary for different locations. To solve this power density variation across the cache module, in this work, we evaluate the dynamic power density on the granularity of each cache bank.

For processor cores, it is difficult to evaluate their power density due to he lack of power simulation tool that is especially targeted to the in-order simple processor cores used in this work. Therefore, in this work, a rough power distribution of the processor cores is assumed based on the prior work by Loh et al. [13] and the simulation tool which is targeted for out-of-order microprocessor [14].

The result of the peak temperature evaluation is shown in Fig. 2.19. In this figure, it shows that the temperature of SRAM stacking chip increases very quickly when the cache size increases, while the temperature of STTRAM stacking chip increase slowly. The major reason is the interaction between leakage power and temperature. The exponential dependence of leakage on temperature implies that the impact of SRAM leakage is more pronounced as the cache size increases. With negligible leakage consumption in STTRAM, its temperature increase due to the increased cache size is much smaller than the SRAM caches. Based on the observation, stacking a large amount of SRAM caches (for example, larger than 16 MB) for future CMPs makes thermal control a big challenge for 3D chips, and replacing SRAM with STTRAM is one of the mitigation solutions.

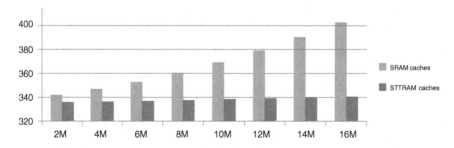

Fig. 2.19 Processor temperatures for different sizes of STTRAM caches and SRAM caches

2.2.5 DRAM Caches Versus STTRAM Caches

Compared to SRAM caches, DRAM–based caches also have advantages of high
density and low leakage power [15, 16]. However, a well know drawback of DRAM
caches is that they need to be refreshed frequently in order to keep data, and refreshing
incurs extra power consumption. In this section, STTRAM and DRAM L2 caches
are compared using our 3D NUCA cache model.

As shown in Table 2.5, the area of one 512 KB DRAM cache bank is similar to that
of one 512 KB STTRAM cache bank. However, the read access latency to DRAM
is much higher than that to STTRAM. In addition, the different of write latency
between DRAM and STTRAM is not as much as that between SRAM and STTRAM.
Therefore, we can expect that the processor can achieve better performance with
STTRAM caches than that with DRAM caches. The normalized simulation results
are shown in Fig. 2.20. The results show that, with our optimization techniques,
replacing DRAM caches with STTRAM caches can improve the performance of
CMPs.

From Table 2.5, we can also find that the leakage power of a DRAM cache bank
is still higher than that of a STTRAM cache bank. Although the access energy to
a DRAM cache bank is lower than that to a STTRAM cache bank, the total power
is reduced if we replace DRAM caches with STTRAM caches because the leakage
power is still dominating. The comparison of total power is shown in Fig. 2.21.

Table 2.5 Parameters comparison between DRAM cache and SRAM caches (65 nm technology)

Cache size	Area (2.38 mm^2)	Read lat. (ns)	Write lat. (ns)	Read ene. (nJ)	Write ene. (nJ)	Leakage power (W)
512 KB DRAM	2.38	4.966	4.966	0.705	0.689	0.2
512 KB STTRAM	3.30	2.318	11.024	0.858	4.997	0.0159

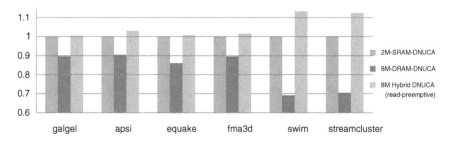

Fig. 2.20 Normalized IPC numbers of using the SRAM/DRAM cache and the hybrid cache

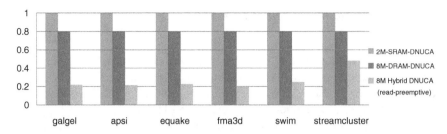

Fig. 2.21 Normalized power of using the SRAM/DRAM cache and the hybrid cache

2.3 Frequent-Value Based PCM Memory

Phase Change Random Access Memory (PRAM) has great potential as the replacement of DRAM as main memory, due to its advantages of high density, non-volatility, fast read speed, and excellent scalability. However, poor endurance and high write energy appear to be the challenges to be tackled before PRAM can be adopted as main memory. In order to mitigate these limitations, prior research focuses on reducing write intensity at the bit level. In this work, we study the data pattern of memory write operations, and explore the frequent-value locality in data written back to main memory. Based on the fact that many data are written to memory repeatedly, an architecture of frequent-value storage is proposed for PRAM memory. It can significantly reduce the write intensity to PRAM memory so that the lifetime is improved and the write energy is reduced. The trade-off between endurance and capacity of PRAM memory is explored for different configurations. After using the frequent-value storage, the endurance of PRAM is improved to about $1.6X$ on average, and the write energy is reduced by 20 %.

2.3.1 Concept of Frequent-Value

The concept of frequent-value locality was first proposed by Zhou et al. [18]. This work pointed out that, at any given point in the program execution, several distinct values occupy a vast fraction of the values in memory accesses. These distinct values are named as frequent values. For some applications, the top three frequently accessed values can occupy more than 50 % of total accesses. Based on this locality, a frequent-value cache (FVC) is proposed. Since only frequent values are kept in FVC, the data in FVC are stored in a compact form. Consequently, FVC can help reduce the cache miss rates efficiently with low overhead.

As frequent-value locality provides the opportunity of storing data in a compact form, it is normally used for data compression [19, 20]. Yang et al. also leveraged the frequent-value locality to reduce the access energy of caches [21]. In their work, frequent values were read and written in an encoded form, the length of which was less than that the original data. Therefore, the dynamic energy consumption is reduced because less bits are activated in the read/write operations.

2.3.2 Frequent-Value Based PRAM Memory

In this section, the frequent-value locality is proved to exist in the data written to memory. Based on this observation, the traditional memory architecture is modified to apply the frequent-value based storage in PRAM memory.

2.3.2.1 Frequent-value Locality in Data Written to Memory

The following terms are defined for the discussion:

- **FV Length**: the bit length of the frequent value.
- **Encoded data**: the data that are identified as frequent values and are stored as encoded form in PRAM memory.
- **Original Data**: the data that are not identified as frequent values.
- **FV Number**: the total number of frequent values.
- **FV Ratio**: $\frac{total\ number\ of\ frequent\ values}{total\ number\ of\ data\ written\ to\ PRAM\ memory}$

In order to identify the frequent-value locality in data written to PRAM memory, diverse applications from SPEC and PARSEC benchmark suites are well studied by using the simulation tool SIMICS. For all benchmarks, the data written back from last level caches are tracked so that the frequently written values can be identified.

The results of FV ratios are shown in Fig. 2.22, in which FV length equals to the size of a cache line (512 Bit), and FV number equals to eight. The results show that more than half of benchmarks exhibit high levels of frequent value locality, indicating that the frequent-value locality exists in the data written to PRAM memory. Specifically, for the several benchmarks, the top eight frequent values occupy more than 40 % of total written data. On average, about 30 % of data written to PRAM memory involve only eight distinct values.

Although frequent-value locality exits in data written to memory, there are several issues should be concerned before applying the frequent-value based storage to PRAM memory: (1) The frequent values need to be stored in a compressed pattern in order to reduce the write intensity to PRAM memory; (2) Although the frequent values are only compressed during write operations, the encoded values need to be identified in read operations; (3) Data compression/decompression may decrease the

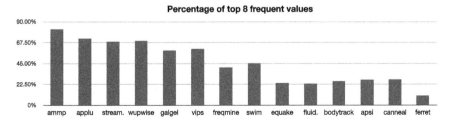

Fig. 2.22 FV ratios for different benchmarks (FV length = 512 Bit, FV number = 8)

performance of write/read operation. The architecture need to be adapted to mitigate the impacts; (4) The frequent values and other parameters, such as FV length and FV number, should be chosen carefully because they may have impacts on the lifetime and performance of PRAM memory; (5) The storage of compressed data induces space overhead in PRAM memory. The trade-off between capacity and lifetime of PRAM memory need to be explored. All of these issues will be discussed in the following subsections and techniques will be proposed to mitigate the overhead.

2.3.2.2 Architecture for Frequent-value Storage

The traditional memory structure need to be modified in order to apply the frequent-value based storage efficiently. Figure 2.23 illustrates a row of memory array in the modified architecture. The memory row is divided into several data blocks based on the FV length. As shown in the lower part of the figure, a data block is composed of data bits and an extra bit called frequent-value-bit (FV-Bit). The data bits in a data block can be used to store either the original data or the encoded data. The FV-bit identifies whether the original data or the encoded data are stored in a data block. When the original data are stored in the data block, the FV-bit is set to bit '0', and all bits are required to represent a valid data, as in the traditional memory. On the contrary, if a frequent value is identified and stored in the data block, only $log_2(FV\ number)$ bits are used to store the encoded data, and the FV-bit is set to bit '1'. It means that only $log_2(FV\ number)$ bits need to be updated during the storage of a frequent value, and therefore the write intensity is greatly reduced.

In addition to FV-Bit inside each data block, an extra status bit, named *Update-Bit*, is also added into each row of PRAM array. The Update-Bit is set to bit '0' when the data is first load from the secondary storage, such as HDD; it is set to bit '1' when data are written back from the last level cache to the row of PRAM array. With the help this Update-bit, the read-only PRAM memory rows are differentiated from those storing updated data. Consequently, read-only PRAM memory can be accessed directly without being involved in the process of dealing with frequent values. The detailed usage of this bit will be shown later in this section.

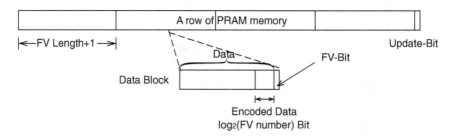

Fig. 2.23 The structure of a row of PRAM memory

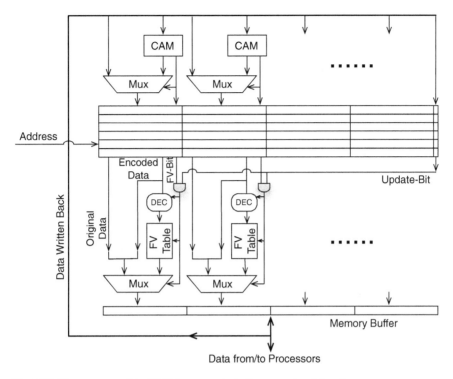

Fig. 2.24 The structure of the PRAM memory including control and data flow

The whole structure view and corresponding data flow of the PRAM memory are shown in Fig. 2.24. The structure includes a memory array, which is composed of PRAM memory rows, and the peripheral circuitry used for read and write operations. In order to simplify the structure, we hide some conventional components, such as address decoders, sense amplifiers, etc., but emphasize the components managing the frequent values.

The circuitry below the PRAM memory array is responsible for read operations. In read operation, the data block is the basic unit of the process. Each data blocks is read out and managed individually from a PRAM memory row. Since either the original data or the data can be stored in a data block, the valid data should be identified before being loaded into the memory buffer. As shown in Fig. 2.24, a decoder (DEC) and a frequent-value-table (FV-Table) are used to translate encoded data to the original data.

The FV-Table records all frequent values, which can be identified in the PRAM memory. The input of the decoder is composed of the $log_2(FV\ number)$ bits used for encoded data. The output of the decoder is used to select the valid data stored in the FV-Table. In order to minimize the timing overhead of read operations, all bits of the data block are read out as original data at the same time when encoded bits are processed. The original data and the output of FV table are fed into a multiplexer

(MUX), which is controlled by FV-Bit and Update-bit shown in Fig. 2.23. Only if both FV-Bit and Update-Bit equal to '1', the decoded frequent value is output from the multiplexer and written into the memory buffer. Otherwise, the original data of the data block are written into the memory buffer.

In order to reduce the energy overhead, FV-Bit and Update-Bit are combined as a signal to make both the decoder and FV table work selectively. Selective working controlled by a signal has already been used in prior research [22]. The value of this signal is just the same as the one used to control the multiplexer, which is shown as the output of the *AND* gate in Fig. 2.24. If a row of PRAM memory is never updated or there are no frequent values stored in the data block, this signal can disable the decoder and FV table.

The write operation is shown as the flow above the memory array in Fig. 2.24. When the data are evicted from the memory buffer and written back to a row of PRAM memory, the data are searched to identify whether they consist of any frequent values. As shown in the figure, the content addressable memory (CAM) is employed to record all frequent values and the corresponding encoded bits. If the data equals to any frequent value in CAM, only $log_2(FVnumber)$ encoded bits are written back to the data block, and both the FV-Bit of the data block and the Update-Bit of the memory row are set to '1'. Otherwise, the original data is written to the data block and the FV-Bit is set to '0'. Note that the CAM and FV-table can be combined in together because they record the same frequent values. They are separated in the figure in order to clearly illustrate the read and write operations.

2.3.2.3 Structure Parameters

The most important parameter is the length of a frequent value because it determines the FV ratios and storage overhead in PRAM memory. As the length of a frequent value decreases, the number of data blocks increases. The storage overhead increases because an FV-Bit is induced in each data block. On the other hand, the FV-Ratio is increased when we use smaller size of data block. It is because the frequent value with shorter FV length has a higher probability to be written repeatedly. The impact of data block size, however, varies for different applications, which have different patterns of data.

Intuitively, we may choose the size of a cache line in the last level cache as the length of frequent value. As we have shown in Fig. 2.22, some applications show high frequent-value locality with such a length of frequent value. The last several applications , however, have very poor FV ratios with such a FV length.

The write intensity, however, may also be increased if the FV length is too small. Although FV-ratio increases with a smaller FV length, more encoded bits may be written to memory because the number of data blocks in a memory row is increased. For example, when FV length is reduced by half, a frequent value is divided into two parts. It is possible that both of these two parts are identified as new frequent values. Consequently, we need $2 \times log_2(FVnumber)$ to store the same data in PRAM memory. When FV length is too small, the increase of encoded bits may offset the benefits from increasing FV-ratio, and total write intensity to PRAM memory may be increased. In addition, the hardware overhead is increased as the FV length is reduced. As shown in Fig. 2.24, the access to each data block is in parallel so that the impact on performance is minimized. Consequently, the number of duplicated peripheral circuitry equals to the number of data block in a PRAM memory line, which is decided by FV length.

FV number also has an impact on performance of the memory architecture. Apparently, a higher FV-ratio can be achieved with a larger FV table because more frequent values can be identified. However, the capacities of FV table and CAM are increased with the FV number. Consequently, the hardware overhead is increased, and it takes more time to decode those encoded $log_2(FV\ number)$ bits and to search the CAM and FV table. It should be mentioned that the benefits of increasing FV number is related to the FV length. For some applications, the frequent-value locality is very low for a large FV length, the FV-ratio may be not increased much with a large FV number.

2.3.3 Profiling and Management of Frequent Values

In this section, two methods of generating frequent values for different cases are introduced. Also, the methods to manage these frequent values for different configurations are discussed.

2.3.3.1 Static Profiling

The static profiling means that the fixed frequent values are profiled on individual application and are generated before running the application. This method is similar to the approach of "*Find Once for a Given Program*" in prior work [23]. This is a software based approach. Since the profiling is finished in advance, it will not cause run-time profiling overhead. In addition, the frequent values are found based on the profiling of the whole program so that the FV ratio can be kept in a high level on average. This method, however, also has its own limitations. It requires the support from compilers, and the profiling of the whole program is time consuming. These frequent values need to be included in the executable code of each application. More importantly, the frequent values can be different with various input data of applications. Therefore, the static profiling is suitable for cases that applications are executed repeatedly with similar input.

2.3.3.2 Dynamic Profiling

Different from the static profiling method, the dynamic profiling method generates frequent values during the run-time execution. It can be achieved in hardware without requiring support from compilers. The run-time profiling overhead, however, may be induced. Some run-time profiling methods are introduced in prior work [23] for on-chip caches or data transaction. In these methods, old frequent values are continuously evicted from the FV table and new values are inserted to the table during the execution. Whenever a frequent value is evicted, a refreshing operation should be called to ensure the consistency in the memory. A refreshing operation is to flush the corresponding frequent value stored in the memory and to restored the original date. However, these methods is not suitable for the PRAM memory because the continuous refreshing operation will cause significant degradation of performance. More importantly, the refreshing operation will induce extra write intensity and therefore reduce the lifetime of the PRAM memory. In order to solve these problems, a novel dynamic profiling technology is proposed for PRAM memory, which generates the frequent value incrementally without causing any refreshing.

The basic idea of this method is to profile the frequent values from the beginning of each program and fill the FV table incrementally during the execution. It is inspired by the observation that *the top frequent values of written data do not vary much during the execution of the application.* This characteristic of frequent-value locality can be proved from results in Fig. 2.25. In the experiments, each application is divided into 10 different stages based on the execution time. The top 32 frequent values of each stage are profiled statically. Then, each application is executed for 10 rounds. In ith round, the top 32 frequent values profiled in ith stage is stored in FV table to work as the frequent values for the whole application. In other words, we examine the efficiency of using frequent values of different stages. The FV ratio of each round is shown in the Fig. 2.25. Note that the straight line is the result of FV ratio of using static profiling. The results show that most of frequent values are common for different stages. Another interesting observation is that *zero is always one of top frequent values for all applications.*

Based on these observations, we propose the method of *global incremental profiling*, which is described as follows:

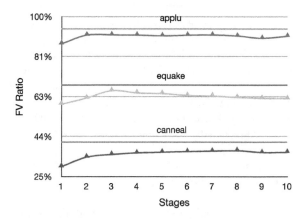

Fig. 2.25 FV ratios using FVs of 10 stages

- The FV table is initialized with only one entry of frequent value, which equals to zero.
- A write access counting table (WAC table) is used to profile the run-time frequent values.
- Each time a data block is written to PRAM memory, the WAC table is updated using the swapping method [23]: If the value in data block is already present in entry i, then the counter at entry i is incremented by one; When the counter saturates, the entry i is swapped with entry $i - 1$ values; When a new value is encountered, and there is no free entry, an entry is freed from the bottom of the table. .
- After a profiling period P_r, the WAC table is compared with the FV table. The value of the first entry in the WAC table, which is not present in current FV table, is inserted into FV table as a new frequent value. The CAM is updated at the same time.
- When the FV table is full, the values in FV table are fixed and used for frequent storage in the rest of execution. The dynamic profiling is disabled by clock gating.

It has been proved that such a swapping method can approximately profile the top frequent values in WAC table without sorting the total access numbers of these values [23]. The WAC table has no impact on the performance because it is not in the critical path. The dynamic profiling quality of WAC is related to the profiling period P_r. This is further discussed in Sect. 2.3.5. Compared to the static profiling, the global incremental profiling is suitable for the case of running applications with different input data.

2.3.3.3 Management of Frequent Values in CMPs

Prior research about frequent-value locality focuses on the architecture in a single-core processor. As the mainstream architecture moves to the modern CMPs, the management of frequent values become more complicated. When the multi-thread/ multi-programmed applications are executed on CMPs, the management of frequent values should be managed carefully to make sure that accesses to frequent values are correct.

In a P-core CMP, it is possible that P different programs are executed at the same on different cores. Consequently, P sets of CAMs and FV tables are required to support co-current accesses of frequent values multiple applications. Since the frequent values are application based, an application ID should be sent together with the memory request in order to access the correct CAMs and FV-tables. The method of sending application ID together with memory request is shown to be feasible in prior research [24]. During the context switch, the values in FV-table of the evicted program should be kept in the PRAM memory for the future access. Then, the FV-table should be refreshed and filled with frequent values of the active program.

2.3.3.4 Wear Leveling of Frequent Values

Wear leveling is still required even when we apply the frequent storage techniques to reduce the write intensity. As shown in Fig. 2.23, $log_2(FV\ number)$ bits are used to store the encoded frequent values. These bits wear faster than the rest in a data block, especially for applications with high frequent locality. Thus, we shift the positions of $log_2(FV\ number)$ bits, which store the encoded frequent values, inside the data block. Note that this shift operation is different from that in prior work [25], because only $log_2(FV\ number)$ bits are shifted.

The FV-bit and Update-bit in the data block, however, are not shifted. After the data is loaded into the PRAM memory line, the Update-bit is, at most, changed once before data is evicted. The FV-bit changes when the value stored in a data block is changed between original data form and the encoded form. The experimental results show that this bit wear our much slower than data cells in the data block. Therefore, the position of these bits are fixed because they will not wear out faster than the data cells.

2.3.4 Complementing with Available Techniques

As we have addressed, our technique of frequent-value storage is to reduce the write intensity to PRAM memory by exploring the locality at data-level. The approaches proposed in prior research focus on the bit-level. It means that the frequent-value storage can be easily integrated with these bit-level techniques to further improve the lifetime of PRAM memory. For example, the DCW technique can be applied with frequent value storage by comparing each encoded bit before writing. For "Flip-N-Write" technique, the only difference is that the hamming distance is just calculated for $log_2(FV\ number)$ bits if a frequent value is written. In addition, the frequent value storage can help reduce the overhead of using these bit-level techniques. For the read operation before write, when an encoded frequent value is identified, only $log_2(FV\ number)$ bits rather than the whole data block need to be read out for comparison.

Similarly, the wear leveling techniques in prior work can also be applied with frequent value storage. Note that, if we apply the shift of frequent values introduced in the previous section, we don't need the byte-level shift operation in prior work [25]. Since a lot of written data can be identified as frequent values, an approximately uniform write intensity can be achieved with the shift of encoded bits for frequent values.

Table 2.6 Baseline configurations

Processor	8-core in-order CMP, 1 GHz
Caches	D/I L1 caches: 32 + 32 KB, L2 caches: 32 MB
PRAM memory	4 GB, 32 KB memory line, 128-entry write buffer read latency: 60 cycles, write latency: 160 cycles per 16 Byte

2.3.5 Evaluations

In this section, comprehensive experimental results are provided and the structure parameters are discussed.

2.3.5.1 Baseline Configuration

The parameters of baseline configuration are listed in Table 2.6. We use an eight-core in-order CMP as the processor. The second level cache has a capacity of 32 MB in order to reduce the write intensity to PRAM memory. The size of write buffer before PRAM memory is set to 128 entries. The write buffer is large enough to mitigate the impact of long write latency of PRAM memory. For all the benchmarks, the write buffer is never full so that the read operation will not be blocked for a long time if there is a burst of write operations. The memory controller, however, does not adopt the technique of "write cancellation" [26] for several reason: (1) The written latency, which varies with the written data, makes it difficult to cancel a write operation; (2) More importantly, the write cancellation aggravate the endurance problem since some write operations are canceled and re-written repeatedly to PRAM memory.

2.3.5.2 The Evaluation Metrics

Applying frequent-value storage to PRAM memory requires extra bits to record the status of a data block or a row PRAM memory. The effective capacity of PRAM is reduced after using the architecture of frequent-value storage. As we discussed, the effective capacity and lifetime of PRAM vary with different configurations of frequent-value storage. It shows the trade-off between the lifetime and capacity of PRAM memory.

In order to explore the trade-off and fairly evaluate the endurance of PRAM memory, we propose a new metric called "capacity-lifetime product" ($Capacity \times Lifetime$). Note that $Capacity$ represents the effective capacity after using techniques to improve the lifetime of PRAM memory.

2.3.5.3 Evaluation of Frequent-value Locality

Figure 2.26 lists the results of FV ratios for different FV numbers and FV lengths. The FV number varies from 8 to 128. The circuit level simulations show that the access latency to CAM or FV table of 128 entries can be finished within one cycle. Consequently, we choose 128 as the largest number of frequent values in this work to minimize the overhead on performance. These frequent values are profiled with static profiling method. The results show that the FV ratio increases when there are more frequent values identified in the PRAM memory. The FV ratios of some applications, however, show low sensitivity to FV numbers. For these applications, there is a very high frequent-value locality in the data written to PRAM memory. FV-ratios are very high even when there are only eight frequent values. On the contrary, the FV-ratios of some applications are sensitive to the number of frequent values, such as *equake*,

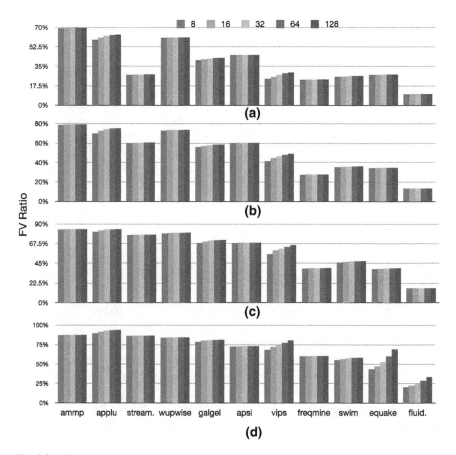

Fig. 2.26 FV ratios for different FV numbers and FV lengths. FV length = **a** 512 Bit; **b** 256 Bit; **c** 128 Bit; **d** 64 Bit

canneal, ferret, etc. Besides the top eight frequent values, other frequent values are also written to PRAM memory intensively.

As we have discussed, the impact of the FV number also depends on the FV length. For the last two applications (*canneal* and *ferret*) in Fig. 2.26a–c, FV-ratios only increase a little, when the FV number is increased to 128. In Fig. 2.26d, when the length of frequent value is reduced to 64 Bit, the FV ratios of these two applications become very sensitive to FV numbers. It is because the length of a register in the processor exactly equals to 64 Bit. Since the data output of the processor is written in such a granularity, a high frequent-value locality is achieved with a FV length of 64 Bit.

The FV ratios for different FV lengths are shown in Fig. 2.27, with a fixed FV number of 128. For all applications, the FV ratios are increased as FV length decreases. For most applications, increase of the FV ratio is not significant when FV length is reduced from 64 to 32 Bit. We also find exceptions in the last several applications. We study the written data in these applications and find that, the upper half of the 64 Bit is kept the same for lots of data from processors. Consequently, the FV ratios are increased much when the FV length is reduced to 32 Bit.

Figure 2.28 compares the FV ratios with different profiling methods. The label *Static* represents the static profiling method; *Global* represents the proposed global incremental profiling method. The incremental profiling period is set to 2^{13} of write operations. For applications with high frequent-value locality, the FV ratios do not vary much for different profiling methods. For applications with low frequent value locality, the frequent values vary a lot in different stages of the program execution. We can achieve larger FV ratios with the static profiling method. Since the static profiling is not suitable for applications with diverse input, in the rest of this work, only the dynamic profiling is employed.

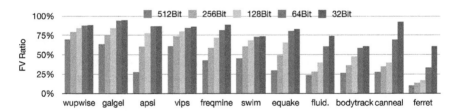

Fig. 2.27 FV ratios for different FV lengths (FV number = 128)

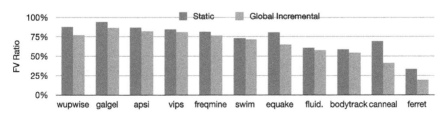

Fig. 2.28 FV ratios of different profiling methods (FV number = 128, FV length = 64 Bit)

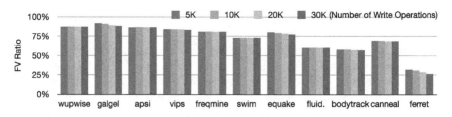

Fig. 2.29 FV ratios of incremental method with different profiling periods (FV number = 128, FV length = 64 Bit)

The results in Fig. 2.28 show that, for most applications, FV ratios of global incremental profiling method are very closed to those of using static profiling. The process of global incremental profiling is finished in the early stage of the program execution. The profiled global frequent values are related to the profiling period, P_r. In Fig. 2.29, we list the results of FV ratios using the global incremental method with different P_rs. For all applications, the FV ratios vary little with P_r. The global frequent values profiled in different stages are similar. Consequently, we can generate similar global frequent values in FV tables with different profiling periods. The FV-ratios, however, decrease a little when the profiling period is too long. The reason is that it takes longer time to fill the FV table with a longer profiling period. Based on these results, we use 2^{13} of write operations as the profiling period. It requires 13 counting bits in each entry of WAC table.

2.3.5.4 Evaluation of Endurance

The improvement of lifetime is evaluated by iteratively running these applications till a failure happens in one PRAM memory cell. Figure 2.30 shows the improvement of lifetime when the global incremental profiling is used. For applications with high frequent-value locality, the lifetime is improved by more than 10 times when the length of frequent value is larger than 64 Bit. For these applications, the improvement of lifetime is reduced when FV length decreases. It is because FV ratios are not increased much with shorter frequent values. However, the number of data blocks in a PRAM memory line is increased and more bits are used to store the encoded bit, as we discuss in Sect. 2.3.2.3. On the contrary, for applications with low frequent-value locality, the highest improvement of lifetime happens when we use 64 Bit or 32 Bit as the FV length. For these applications, the FV ratios are greatly increased with short frequent values and we have more benefits of storing more frequent values as encoded bits.

In Figure 2.31, we evaluate the benefits of combining data-level and bit-level techniques together. We compare the results of $Capacity \times Lifetime$ between two cases. In the first case (DCW), only the bit-level technique in prior work is used. In the second case ($DCW + FV$), the frequent value storage is used in together with DCW. The results show that the $Capacity \times Lifetime$ can be improved to about

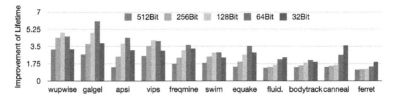

Fig. 2.30 Lifetime improvement with the global incremental profiling

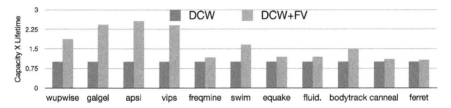

Fig. 2.31 Lifetime improvement of the combined technology

3X in the best case. On average, the $Capacity \times Lifetime$ of using both techniques is improved to $1.6X$ of that only using only bit-level technique.

2.3.5.5 Evaluation of Write Energy

Figure 2.32 compare the results of write energy for different applications. The first case is using the baseline PRAM memory without using frequent value storage. The second case is using frequent-value storage with global incremental profiling. The P_r is set to 2^{13}. The FV length and FV number are set to 128 Bit and 64, respectively. The results show that, on average, the write energy is reduced to 80 % of the baseline after using the frequent-value storage.

2.4 Hybrid SSD Using NAND-Flash and PCM

In recent years, many systems have employed NAND flash memory as storage devices because of its advantages of higher performance (compared to the traditional hard disk drive), high-density, random-access, increasing capacity, and falling cost. On the other hand, the performance of NAND flash memory is limited by its "erase-before-write" requirement. Log-based structures have been used to alleviate this problem by writing updated data to the clean space. Prior log-based methods, however, cannot avoid excessive erase operations when there are frequent updates, which quickly consume free pages, especially when some data are updated repeatedly.

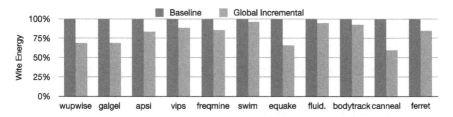

Fig. 2.32 Reduction of write energy

In this section, we propose a hybrid architecture for the NAND flash memory storage, of which the log region is implemented using phase change random access memory (PRAM). Compared to traditional log-based architectures, it has the following advantages: (1) the PRAM log region allows in-place updating so that it significantly improves the usage efficiency of log pages by eliminating out-of-date log records; (2) it greatly reduces the traffic of reading from the NAND flash memory storage since the size of logs loaded for the read operation is decreased; (3) the energy consumption of the storage system is reduced as the overhead of writing and reading log data is decreased with the PRAM log region; (4) the lifetime of NAND flash memory is increased because the number of erase operations are reduced. To facilitate the PRAM log region, we propose several management policies. The simulation results show that our proposed methods can substantially improve the performance, energy consumption, and lifetime of the NAND flash memory storage.

2.4.1 Background

In this section, we first illustrate the erase-before-write limitation for NAND flash memory. Then, we describe the existing IPL method and its limitation. Finally, we present a brief overview of PRAM technology and the potential to solve the limitation of IPL method.

2.4.1.1 The Erase-Before-Write Limitation

NAND flash memory shows asymmetry in how they read and write. While we can read any of the pages of a NAND flash memory, we must perform an erase operation before writing data to a page. The erase operation is performed in the granularity of an "erase unit" consisting multiple adjacent pages [27, 28]. Therefore, writing new data to the same page storing the old data, namely "in-place" update, cannot be performed directly. Instead, it is finished in several steps: (1) back up the other pages in the same erase unit; (2) perform an erase operation; (3) write both the new data and the backed up pages to the erase unit. Such a process is very time consuming.

Because of this erase-before-write limitation, log-based file systems are commonly used for flash memory devices. When some data in a page is updated, the whole page holding the data item is written to another erased page, which is called the log page of current page. Then, the old copy of page is invalidated. This process is called an "out-of-place" updating. The merge operations will happen when the device may run out clean pages, and some invalid pages need to be reclaimed. This process is also called *garbage collection*. During the garbage collection, valid pages in these two erase units need to be merged into a third erase unit. This merge operation is very costly because all valid pages of the erase unit should be written to another erased unit.

2.4.1.2 IPL Method

It is easy to find that even if only a small portion of a page is updated, the whole page must be written to its log page in the log-based file systems, which can significantly degrade the write performance of the file system. Unfortunately, small-to-moderate sized writes are quite a common access pattern for many cases including database applications such as OLTP [28, 29].

Recently, the IPL method [28] is proposed to overcome such a weakness of the log-based file systems for NAND flash memory. Figure 2.33 gives an illustration of the IPL method for NAND flash memory. Each erase unit consists of 60 data pages and 4 log pages. Each log page is divided into 16 *log sectors*. Unlike the traditional log-based file system, the log page in IPL method is used to record updates of data pages rather than copy a whole data page directly. Whenever an update is performed on a data page in the DRAM data buffer, the copy of data item in the buffer is updated in-place. At the same time, the IPL buffer manager adds a log record in the data buffer, which is allocated in the log sector assigned to the data page. A log sector in the data buffer can be allocated to the data page on demand, and it can contain more than one update record. A log sector in the data buffer can be released when its log records are written back to the log sector in flash memory. The log records are written to NAND flash memory when the log pages of an erase unit become full or when a dirty data page is evicted from the data buffer. When a dirty page is evicted, the data page itself does not need to be written back to flash memory, because all of its updates are stored in the form of log records. Consequently, only its log sectors are written back to NAND flash memory.

Since the number of the log sectors is fixed for each erase unit, the erase unit may eventually run out of empty log sectors after certain number of updates are performed to the pages in the erase unit. For such a case, the IPL manager performs a merge operation as follows. First, it merges the old data pages with their log sectors to create up-to-date data pages. Second, it allocates a clean erase unit to write the generated up-to-date data pages. Finally, it invalidates all the pages in the old erase unit, so this erase unit can be reclaimed for future use.

In the IPL approach, only the log sectors that contain the updated data are written back to flash memory, and the usage of data pages is more efficient than that of

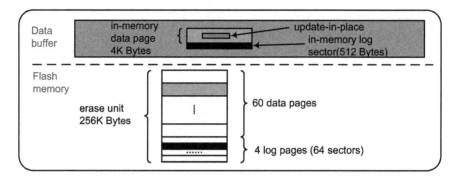

Fig. 2.33 An illustration of the IPL method

traditional log-based flash memory. The size of log pages in an erase unit is limited, and these log sectors themselves do not allow in-place updating. This may cause significant performance degradation for some cases. Particularly, if there are frequent updates to the same erase unit, it would quickly run out its log sectors and cause merge operations frequently. Moreover, if there are multiple updates to the same data, only the latest updated one is valid. Consequently, the effective log sector capacity of an erase unit becomes smaller, which would worsen the problem of the IPL approach.

2.4.1.3 PRAM Versus NAND-Flash

As PRAM technologies improve, PRAM has shown more potential to replace NAND flash memory with advantages of allowing in-place updates and fast access speed. Table 2.7 compares the characteristics of PRAM and NAND flash memory, which are estimated from prior work [30–32]. Note that the units of write and read operations for PRAM and NAND flash memory are different. The PRAM can be accessed in a fine granularity (byte-based). We will show that this advantage makes the access to the log region much more flexible, compared to the traditional IPL method. Prior research has also shown that the PRAM could achieve the same cell size as that of NAND flash memory with a vertically stacked memory element over the selection device [33, 34]. It means that it is feasible to replace NAND flash memory with PRAM without inducing area overhead.

Currently, it is still not feasible to replace the whole NAND flash memory with PRAM entirely due to its high cost and the limitation of manufacture [33, 35]. Consequently, we propose to use the PRAM as the log region of NAND flash memory instead. A cell of NAND flash memory could be implemented to store multiple bits of data. The multi-level PRAM is still under research [36]. In this work, the single-bit PRAM is used.

Table 2.7 The comparison between PRAM and NAND flash memory technologies

Tech.	Cell size (F^2)	Write cycles	Access time			Access energy		
			Read	Write	Erase	Read	Write	Erase
Flash	4	10^5	284 μs/ 4 KB	1833 μs/ 4 KB	>20 ms/ Unit	9.5 μJ/ 4 KB	76.1 μJ/ 4 KB	16.5 μJ/ 4 KB
PRAM	4	10^8	80 ns/ word	10 μs/ word	N/A	0.05 nJ/ word	0.094 nJ/ word	N/A

2.4.2 Overview of the Hybrid Architecture

Figure 2.34 shows the physical and structural views of NAND flash memory storage system with the PRAM-based log region. Since the process technologies of PRAM and NAND flash memory are different, it is difficult to physically place the PRAM-based log region together with the flash-based data region. Unlike the IPL method, the PRAM log region is separated from the flash-based data region in our work as shown in Fig. 2.34a. Based on log region management policies and run-time operations of applications, log sectors are dynamically assigned to each erase unit to store its own updates. The structural view of such relationship is shown in Fig. 2.34b. We need to maintain metadata for the relationship between data pages and their corresponding log sectors. The *log region controller* in Fig. 2.34 takes the responsibility of decoding addresses and managing the metadata for the log region. The access to the hybrid architecture is described as follows:

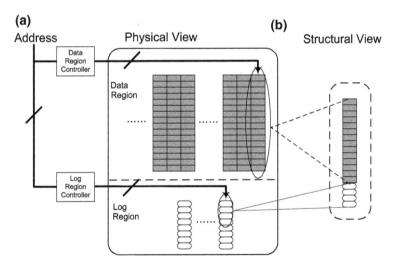

Fig. 2.34 Physical and structural views of the hybrid architecture

- For the read operation, the address of the accessed data is sent to both the data region and the log region. If there exist log sectors for the requested data page, they are loaded into the data buffer as well as the original data page to create the up-to-date data page.
- For the write operation, only updates are sent back to the PRAM log region. There are several scenarios:

  - Case1: If there are no existing log records for the accessed data page, a new log sector is allocated to the accessed data page. Then, the current update is written to the log sector.
  - Case 2: When some log sectors have already been allocated to the accessed data page, the log records in these log sectors are compared with the current update. If there is a log record that has the same data address of the current update, the existing log record is overwritten by the current update, and no extra log record is generated.
  - Case 3: If the current update does not match any of existing log records, a new log record containing the current update is written to the log sectors of the data page. When log sectors assigned to the data page are fully occupied, a new log sector is requested for the current update as in Case 1.

- Merge operations are triggered when the merge conditions are satisfied. The merge conditions depend on the log region management policies, which will be introduced in following sections. During merge operations, updates in these log sectors are applied to corresponding data pages, and the data pages are written to another free erase unit. The out-of-date data pages are invalidated and the corresponding log sectors are released as clean ones for future use.

The data address of the current update needs to be compared to those of existing log records in order to achieve the in-place update. A relative address ($addr_{Relative}$) is used to represent the position of an update inside the accessed data page. It can be calculated by $addr_{Relative} = addr_{Update} - addr_{Page}$. The $addr_{Update}$ and $addr_{Page}$ represent the real addresses of the update and the accessed data page, respectively.

As we mentioned, an access to a PRAM log region is managed with its own controller. The access to the log region is operated in parallel with that to the data region. Since the size of a log region is much smaller than that of a data region, the accessing delay on the peripheral circuitry of the log region is shorter than that to the data region. It means that using hybrid architecture will not induce extra delay. Instead, the total access latency could even be reduced in some scenarios. For example, in write operations, the latency is decided by the time of writing the update to the log region. Then, the performance could be improved with a shorter decoding time.

Besides the shorter access latency, the more important thing is that the hybrid architecture can take advantages of in-place updating capability of the PRAM log region. Although it takes extra time to search whether the current update does not match existing log records, the overhead is trivial compared to the write latency

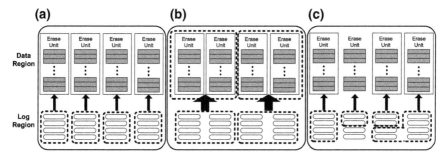

Fig. 2.35 Examples of **a** the basic static log assignment, **b** the group static log assignment, **c** the dynamic static log assignment

because the read operation is much faster than the write one. If the in-place updating happens, much more benefits can be achieved. For example, since the update is written to the existing record, the time of allocating new log record and modifying metadata is saved. It is also beneficial in that the log region is used more efficiently: it would reduce the numbers of erase and write operations by increasing the effective capacity of the log region.

Furthermore, because the PRAM log region can be accessed with byte granularity, the performance of read operations can also be improved. In the IPL method, since the log data is loaded in the page granularity, some log sectors that do not belong to the accessed data page are also loaded. If the log records of one data page are stored in more than one physical page of NAND flash memory, the overhead of loading the log records can be even higher than that of loading the data page. On the contrary, in our hybrid architecture, only the log records that belong to the accessed data page are loaded. It can greatly enhance the performance of read operations.

2.4.3 Management Policy of PRAM Log Region

The in-place updating and fine access granularity of PRAM log region provide opportunities of optimization. In this section, we propose a few assignment and corresponding management policies for the PRAM log region.

2.4.3.1 Static Log Region Assignment

In the IPL method, each erase unit is uniformly assigned a fixed number of log pages. It is obvious that such a static assignment also works with the PRAM log region. This method of assignment is named as the *basic static assignment*. Its illustration is shown in Fig. 2.35a. When the data pages in the erase unit are updated, only the log sectors that belong to the erase unit are assigned for these updates. Although the static log sector assignment is the same as that in IPL method, the performance can be improved using the in-place update capability of PRAM. As in the IPL method, the merge operation is triggered when all log pages of an erase unit are fully occupied and no free log sector can be assigned to a new update. Note that the new update means that the incoming update is not matched to any existing log records. With the PRAM log region, when all log pages of an erase unit are used, it is possible that the logs are written in-place without causing merge operations.

The main advantage of the static assignment is its simplicity in implementation. Since the log sectors assigned to each erase unit are predetermined and fixed, the overhead of keeping and searching the metadata of log sectors is negligible. Such uniform assignment, however, becomes inefficient when updates are not evenly distributed among erase units. In applications such as OLTP, the access/update intensity to each erase unit is typically not uniform. Normally, the distribution of update numbers in respect of erase unit is highly skewed for a trace of TPC-C benchmark. Note that TPC-C is a popular benchmark that simulates a complete computing environment where a population of users executes transactions against a database [37]. Thus, for such applications, it is inefficient to divide and assign the log region equally to each erase unit.

One way to mitigate the effect of uneven access distribution is to organize erase units into groups, which is named *group static assignment*. In this way, log pages in a group can be shared among erase units, as shown in Fig. 2.35b. After sharing log pages in a group, the frequently updated erase units in a group could be assigned more log pages. Note that such a method of sharing log pages among erase units is not feasible for the IPL method using only NAND flash memory. It is because the log pages are placed together with data pages inside each erase unit of the IPL method. If we want to assign data pages with external log pages from other erase units, a pointer-like structure is needed to record locations of external log pages, which may increase the design complexity and timing overhead. Furthermore, since the log region is managed in page granularity, sharing log pages among erase units may generate some data pages, which have too many log pages. On the contrary, in our PRAM log region, all log pages are placed together. There is no difference whether we manage log pages in the granularity of an erase unit or in a group with several erase units. Similar to the basic static assignment, the merge operation is also triggered when the log pages assigned to each group run out. Note that all erase units in the group need to be merged together.

Although the grouping method could help in mitigating the effect of unbalanced accesses, there are some limitations with it. First, if most of erase units inside a group are frequently updated, the log pages shared in this group cannot reduce the number of merge operations much. Second, it is possible that only a few erase units in a group are frequently updated and cause many merge operations. Since the whole group are erased and written together in the merge operation, most pages are erased even without any modifications. Therefore, the size of a group could not be too large in order to avoid the overhead of such redundant erases. Third, it takes more time to access and manage log pages as the size of a group increases.

2.4.3.2 Dynamic Log Region Assignment

In order to overcome the limitations of the static assignment, we propose a dynamic log page allocation method. The basic idea of the dynamic allocation method is that the number of log sectors of an erase unit could be assigned on-demand based on the number of its updates. In other words, the log region is *shared* among all erase units. If an erase unit is frequently updated, the log region controller assigns more log sectors to it unless there is no remained free log sectors in the log region. Note that such a dynamic assignment is different from the group static assignment. The log sectors of each erase unit are managed individually, and the merge operation of one erase unit will not affect the others.

Figure 2.35c gives an illustration of the dynamic assignment. At the beginning, there are no updates to any data page, and all log sectors are free. During the access process, the number of log sectors assigned to each erase unit grows up based on its updates. The frequently updated erase unit are assigned more log sectors.

The dynamic assignment promises that frequently updated erase units can get more log pages so that the number of merge operations are significantly reduced for applications having unbalanced updates. Such a dynamic assignment is not feasible for the IPL method with the NAND flash memory log region, which does not allow in-place updating. It is obvious that the number of the log sectors is increased linearly in proportion to the size of updates when the in-place updating is not allowed. When the updates to erase units are unbalanced, the frequently updated erase unit may have too many log sectors, and the overhead of accessing such erase unit is not tolerable. On the contrary, in our PRAM log region, the number of log sectors is not increased when some data are repeatedly updated. In the worst case, the size of logs is the same as that of the data. Therefore, accessing overhead for the dynamic assignment is moderate with the PRAM log region.

Intuitively, the merge operations are triggered when the whole PRAM log region is fully occupied. In that case, however, once the merge operation is triggered, all log sectors should be merged to the corresponding data pages at the same time, which makes the storage system not available for a long time. Hence, if the burst of updates arrives when the log region is nearly full, the whole system has to be stalled for a long time waiting for merge operations to be completed, which is not desirable. Instead, we set up a threshold of free log sectors, and the merge operations are triggered when the capacity of free log sectors are lower than it. Its advantage is that there are always some free log sectors reserved for the burst of updates. In addition, it is more likely to process merge operations during the idle time of the storage system.

With the dynamic assignment method, the whole log region is managed together. It is possible to reduce the number of merge operations by merging eras units selectively instead of merging them all together. Such policies are described as follows,

- If some erase units are just frequently updated, it is highly possible that these pages will be accessed repeatedly in the near future. Such erase units are named as *hot units* in this work. If we could prevent these hot pages from being merged, the number of merge operations could be reduced. A simple FIFO queue is employed to record most recently updated erase unit. When the merge operation is triggered, erase units in the queue are left untouched so that hot data will not be merged.

- The data of whole erase unit is copied to a clean space, and the timing overhead mainly depends on the erase time. Therefore, it is not worth merging the pages that have only a few log sectors. In another word, merging such erase units is not efficient because it will not releases many log sectors. Consequently, we could set up a threshold based on the number of log sectors of the erase unit, which is named as *merge threshold*. The erase unit is kept untouched during the merge process if the number of its log sectors are lower than the merge threshold.

The hot unit queue size and the threshold of log sector number have the important impact on the number of merge operations and the lifetime of the system. These issues are further discussed in the experimental sections. The size of metadata for dynamic assignment is larger than that for static assignment. Since we want to share the whole log region among data pages, we need more bits in metadata to store the location (address) of the log sectors. In this work, the metadata is located in the PRAM because it is normally frequently accessed and modified [38]. It means that the available log region capacity is reduced by storing metadata. In addition, we need to keep the record of hot erase units with a FIFO queue. In this work, a link based table structure is used for keeping the metadata of log sectors. Note that some data structures, such as the hash table, could be employed to reduce the space and timing complexity of searching and accessing the metadata. Such topics are out of the scope of this work and are not discussed. We will discuss overhead of storing metadata later, and results show that the performance is still improved with the overhead.

2.4.4 Endurance of the Hybrid Architecture

The lifetime endurance is one important issue of the NAND flash memory. There are various good approaches proposed for wear leveling of NAND flash memory [39–41]. When the in-place updating is enabled in the hybrid architecture, the endurance of the storage system can be improved. Since updates are written to the PRAM-based log region, the write intensity to the data region is greatly reduced compared to the pure NAND-flash memory. For the same applications, the lifetime of the data region is increased with the PRAM-based log region. Because PRAM has much better endurance than the NAND flash memory, the log region may still wear slower than the data region does. If we promise that the log region will not wear out before the data region, the endurance of the whole system is increased. As we mentioned in the previous section, we can use selective merge operations to reduce the number of merge operations. The lifetime of the data region is further improved at the same time, but the log region may wear faster. It is a trade-off between performance and the endurance of the log region. Consequently, the wear-leveling is necessary for the PRAM-based log region. In this section, we discuss the issues related to the lifetime of the log region and propose the technique for wear-leveling.

2.4.4.1 Lifetime of the Log Region

In the IPL method, the log pages of an erase unit can only be written once before being merged with data pages. Therefore, log and data pages have the same level of wearing. On the contrary, in our PRAM hybrid system, the merge operations should be controlled to promise that the log region will not wear out before the data region. Besides the basic merge conditions introduced in the previous section, several other merge conditions are enforced to prevent the log region from wearing too fast.

For the basic static log assignment, since the log sectors assigned to each erase-unit are fixed, the lifetime of each erase unit could be controlled individually. Table 2.7 shows that the number of allowed writes of PRAM is about 1000 times larger than that of NAND flash memory. Theoretically, if there are 1000 updates to the same log record, the erase unit should be merged. Then, the data and log regions have the same wear level. In this work, a threshold is set up for each log sector. If a log sector is updated up to 1000 times, a merge operation is triggered for the erase unit. In the worst case, a single log record in the log sector will not be updated for more than 1000 times before the erase unit is merged. Therefore, the log region will not wear out faster than the data region does. Such method also works with the group static log assignment. In fact, even for applications with high intensive write operations, such case of forced merge operations rarely happens. With the static assignment, the lifetime of the hybrid system is decided by the data region. Since the number of merge operations is reduced with the hybrid architecture, the endurance is also improved. Note that a 10-bit counter is needed for each log sector.

For the dynamic log assignment, the case is more complicated because the log region is shared among all data pages. It is possible that some log sectors are reused repeatedly by data pages with intensive updates. Therefore, using a threshold as in the static assignment may not efficiently prevent the log region wearing out before the data region does. In addition, the lifetime of the system is related to the management policies. First, the number of merge operations could be reduced if the policy prevents the hot pages being merged. The log sectors belonged to these hot pages may wear faster because they may be used for a long time without being merged. Second, the number of merge operations could be reduced if only data pages with large number of log sectors are merged. It is possible that some data pages are updated for a few times, then, they are not accessed for a long time, which are called *cold pages*. Since these cold pages would not be merged for a long time, their log sectors will not also be used for a long time, which contributes to the unbalanced wear-leveling. We introduce the techniques of wear-leveling to deal with these problems and improve the lifetime of the log region.

2.4.4.2 Wear Leveling

With the dynamic log assignment, when a data page requires for a new log sector, the controller chooses one from the pool of free log sectors. The write intensities of log sectors may be unbalanced with random assignments of log sectors. Therefore, for each request, the log sector with the lowest write numbers should be assigned to even out the write intensities. Theoretically, we should keep tracing the total number of write operations for each log sector. As the PRAM can normally endure 10^8 write cycles, it needs a 30-bit counter to record the total number of write operations. In order to reduce the area and timing overhead, we can use a small counter instead and reset its value periodically to track the approximate wear-level of each log sector. As we introduce in Sect. 2.4.4.1, a 10-bit counter is needed for each log sector in the static log assignment. This counter can be employed in the dynamic log assignment for wear leveling. A wear-aware dynamic log assignment is described as follows:

- The free log sector pool is organized as a link structure. Initially, all log sectors are randomly linked together.
- For each requirement, the head sector of the link is evicted and assigned.
- After each merge operation, the released log sectors are inserted into the link structure based on the number of write operations recorded in its counter. The link structure is kept in sorted order according to the number of writes of each log sector.
- When any of the counter reaches the maximum counting number, the write number of the log sector at the tail of the link structure is subtracted from all counters.

With this structure, the free log sectors are sorted approximately based on their write cycles. The write intensities are balanced among log sectors.

The management of log assignments also incurs timing overhead. For the read operation, the latency of accessing the metadata (record entries) varies for each page. With the in-place updating ability, the log size of a data page is no larger than that of one data page. Therefore, it takes less than 100 μs to access all log record entries for each data page. For the write operation, the metadata need to be updated if there is a new log sector assigned to a data page. In the worst case, the whole log sector is occupied with the new updates, and a new log sector is assigned to the data page. In this case, the size of updated PRAM log region is the same as that in the NAND flash log region. Since the write latency to PRAM is less than that to NAND flash memory, the latency of this part is reduced.

During the merge operation, it takes less time to release log pages in PRAM log region than that in NAND flash one, because the average log size of a data page is reduced with the PRAM log region. In order to achieve the wear-aware log assignment, it takes extra latency to update the sorted tree of free log sector. In the worst case, the timing complexity of searching the binary tree is less than 100 μs, which is trivial compared to that of a merge operation (>20 ms).

2.4.5 Experimental Results

In this section, we presents the simulation results of our hybrid architecture with different configurations. Then, the results are analyzed and compared with those of prior work.

2.4.5.1 Experimental Setup

We use TPC-C benchmarks [37] for the evaluations, which are generated using an open source tool *Hammerora* [42]. The tool runs with a mysql database server on a Linux platform under different configurations. In our experiments, the configuration of transactions in these benchmarks are described in the Table 2.8. Note that different sizes of data buffers are simulated. The current operating systems do not support the management of data buffer as shown in Fig. 2.33. In order to obtain proper log-based access traces to NAND flash memory, we implemented a simulation tool of data buffer with the structure shown in Fig. 2.33. This simulation tool uses memory requests from the processor to the database as the input. Then, the accesses to NAND flash memory storage system are generated based on these memory requests with the log-based data buffer. In order to obtain quantitative evaluations of performance, power consumption and lifetime, we implement the models of PRAM and corresponding simulation tools in device, circuit and system levels. For the NAND flash memory, we study recent work and products specifications from industries [28, 31, 32, 43], then, we adjust and integrate proper parameters into our simulation tools.

The sizes of a page, an erase unit, and a log sector are set to be 4 KB, 256 KB, and 512 Bytes, respectively. For the configuration of the IPL method using only NAND flash memory, there are four log pages in each erase unit. We have mentioned in Sect. 2.4.1 that if the multi-level storage is supported with the PRAM log region, the area of log region is not increased if we replace NAND flash memory with PRAM of the same capacity. In order to explore the advantages of using PRAM, the size of PRAM log region is reduced to be half of the flash log region. Therefore, the area of log region is still kept the same even if the multi-level storage is not employed for the PRAM log region. We will show that performance is still improved. Thus, for the simple static assignment configuration of hybrid architecture, there are 8 KB of PRAM logs in each erase unit. For the dynamic assignment, the merge operations are triggered when the size of free log sectors are lower than 30 % of the total size of PRAM. Then, we ensure that the system will not be stalled for a long time when burst write operations happen.

2.4.5.2 Write Performance Simulation

As we addressed before, the large overhead of write and erase operations is the obstacle of improving performance of NAND flash memory. Our main goal of using the PRAM log region is to improve the write performance. Similar to prior work, we also consider the impact of the data buffer size on the write performance. As shown

Table 2.8 Configurations of benchmarks

1G.(20–100)M.100u	1 GB database, 100 simulated users, the size of buffer pool varies from 20 to 100 MB

in Fig. 2.36a, the number of write operations is reduced as the size of data buffer increases. Note that the different buffer sizes are considered to show the efficiency of PRAM log region under different environments. The actual buffer size may be decided in real cases for different applications.

The comparison of merge numbers with different methods is shown in Fig. 2.36b. The *IPL* represents the pure NAND flash memory using the IPL method. The *s-PRAM* represents the hybrid architecture using the basic static log assignment. The *d-PRAM* represents the hybrid architecture using the basic dynamic log assignment. It means that there are no hot unit queue or merge threshold for optimization of merge operations.

The results show that the number of merge operations is reduced when the in-place updating is enabled in the PRAM log region. For the static log assignment, the number of merge operations is reduced by 22.9 % on average, compared to that of the IPL method. We can find that the hybrid architecture works better when the size of data buffer is smaller. The reason is that, some write operations to the storage system may be filtered when the size of data buffer is increased. Then, the write intensities become more unbalanced. Therefore, the static log assignment method works less efficient (as discussed in Sect. 2.4.2). We can also find that using the dynamic assignment of log pages further reduces the number of merge operations significantly. On average, the number of merge operations is reduced by 58.5 %, compared to the IPL method. This observation is consistent with the previous discussion. All log sectors are assigned on demand with the dynamic assignment. Consequently, we get more benefits from the in-placing updating with the dynamic assignment of log pages. The more important thing is that the size of data buffer has little impact on the working efficiency of the dynamic assignment method. As we discussed before, the dynamic assignment method works well even when the write intensities are highly unbalanced.

The write time is also simulated and compared in Fig. 2.36c. The similar conclusion can be drawn for the write time because the results have the same trend as that of merge operations. It is reasonable because the time consumed by merge operations is dominating in total time of write operations.

In Fig. 2.37a, the numbers of merge operations are shown for the hybrid architecture using the group static log assignment. The results include the numbers of merge operations with different group sizes. The number of merge operations decreases as

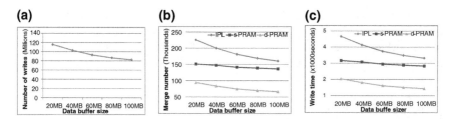

Fig. 2.36 a The impact of data buffer size on the write operations; **b** The comparison of merge numbers; **c** The comparison of write time

the group size increases, but the improvement is not significant. We have mentioned that the overhead of management is increased with the size of group. The results show that we can get more benefits from the group method when the size of data buffer is larger. It is because the group method is used to mitigate the effect of unbalanced write intensities, and the write intensities become more unbalanced with a larger size of data buffer.

In Fig. 2.37b, the results of using different sizes of hot unit queues are compared. At the beginning, the number of merge operations decreases as the size of hot queue increases. It means that we can get more benefits from in-place updating of hot erase units, if they are not merged. The number of merge operations is increased when the queue size is too large. It is because too many erase units are kept untouched during the process of merge operations. The capacity of free log sectors is reduced greatly. Furthermore, many erase units in the queue are not hot ones when the queue size is too large. The results show that the method works best with the queue size of 16. The number of merge operations is reduced by 6.9 % on average.

Figure 2.37c shows the results after we set up the threshold number of log sectors for merge operations. The number of merge operations could be reduced significantly after we set up the threshold because the log sectors are released more efficiently for each merge operation. Similarly, the threshold cannot be too large. Otherwise, the total number of free log sectors are reduced greatly, and the number of merge operations is increased. The results show that the method works best with the merge threshold of 128. The number of merge operations is reduced by 35.2 % on average.

2.4.5.3 Read Performance Simulation

For different methods, the sizes of data and log that are read from the NAND flash memory are listed in Table 2.9. The *g-PRAM-4* represents the hybrid architecture using the group static log assignment, and there are 4 erase units in each group. The *d-PRAM-16-128* indicates the hybrid architecture, which uses the dynamic log assignment with the hot unit queue and merge threshold. The size of hot unit queue is set to 16, and the merge threshold is set to 128. The *Average Overhead* shows the

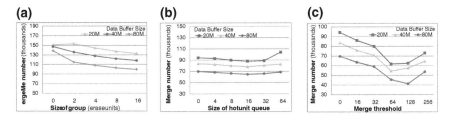

Fig. 2.37 a The impact of the group size on the group static assignment. **b** The impact of the hot unit queue size on the dynamic assignment. **c** The impact of the merge threshold on the dynamic assignment

average percentage of log data in total data per read operation. The results of static and dynamic assignments are compared to the IPL method. The *Reduction* represents the decrease of overhead for reading log data in each read operation, when results are compared to the IPL ones.

The results show that the average time of a read operation is also reduced with the PRAM log region because of two reasons. First, for each read operation, the number of loaded log records is reduced with the in-place updating. In addition, the read speed of PRAM is faster than that of NAND flash memory. The effect of the data buffer is also considered for the read operation. The three sets of results show that we can get similar benefit from using the PRAM log pages with different sizes of data buffers. One interesting observation is that more log pages are read if we use the dynamic assignment method. It is because the frequently accessed erase units are assigned more log pages. The frequent updates also generate more log pages for the same data page. And the numbers of read operations to these erase units are much larger than those of other ones because of intensive accesses. Consequently, the total number of read operations are increased. Nevertheless, it is worth using the dynamic allocation method because the write performance is increased greatly. With write and read evaluation results, the estimated total execution time is compared for managements policies, which is shown in Fig. 2.39a.

2.4.5.4 Energy Consumption Evaluation

The energy consumption for different methods is compared in Fig. 2.38. The write operations energy is shown in Fig. 2.38a. The dynamic log assignment consumes the least energy for the same benchmark because the number of merge operations is greatly decreased with this method. Furthermore, the total size of log written to the log region is reduced with the ability of in-place updating. For the read operation, using

Table 2.9 Read performance evaluations

Methods	Log records read	Avg. overhead (%)	Reduction (%)
1 GB data, 20 MB data buffer, Pages Read:103220			
IPL	6178 pages	5.99	–
g-PRAM-4	23680 KB	2.87	52.1
d-PRAM-16-128	33320 KB	4.04	32.6
1 GB data, 40 MB data buffer, Pages Read: 87600			
IPL	5715 pages	6.52	–
g-PRAM-4	21040 KB	3.00	54.0
d-PRAM-16-128	29768 KB	4.25	34.9
1 GB data, 80 MB data buffer, Pages Read: 68240			
IPL	4742 pages	6.95	–
g-PRAM-4	18080 KB	3.31	52.3
d-PRAM-16-128	24304 KB	4.45	35.9

PRAM log region can also help to reduce the energy for two reasons. First, the average size of log data loaded with one read operation is reduced. Second, it takes less energy to access the PRAM than that to access the same size of NAND flash memory. Note that the dynamic assignment method consumes a little more read operation energy than that of the static method. It is also because the frequently accessed erase units are assigned more log pages. The total energy consumption is also compared in Fig. 2.38c. Since the energy of write and merge operations dominates, the dynamic log assignment still consumes the least energy consumption when both read and write operations are considered.

2.4.5.5 Lifetime Evaluation

In this work, we assume the write cycles of NAND flash memory are 10^5, and the write cycles of PRAM are 10^8. In order to evaluate the lifetime of the data region, the benchmark traces are kept feeding into the simulation tool. We keep tracking the number of write operations till one cell of data region wears out. The lifetime of the log region is simulated in the same way. Then, based on the traces, we can estimate and compare the lifetime.

The lifetime of data and log regions with the best dynamic assignment(d-PRAM-128-6) are compared in Fig. 2.39b. The first column is the lifetime of the data region. The second column is the lifetime of the log region without using any wear leveling technique. The third column is the optimized lifetime of the log region with the wear leveling technique. We can find that, without wear leveling, the lifetime of the log region is just a little longer than that of the data region for benchmarks in this work. It is possible that the log region wears out first for benchmarks with a little higher write intensities. After we use the wear leveling technique, the lifetime of the log region is increased to about 10 times higher than that of the data region. It promises that the log region will not wear out before the data region does, even for benchmarks with much higher write intensities. Then, the lifetime of the whole storage system is decided by that of the data region. The lifetime of the whole storage system are compared in Fig. 2.39c with different configurations. The results show that the lifetime of the whole storage system is improved after using the PRAM log region. It is because the

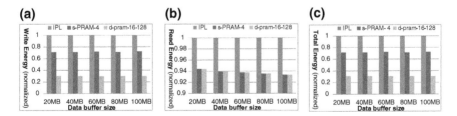

Fig. 2.38 **a** The comparison of write operation energy; **b** The comparison of read operation energy; **c** The comparison of total energy consumption

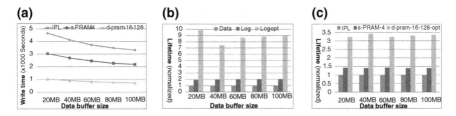

Fig. 2.39　**a** The comparison of the total execution time; **b** The comparison of lifetime between the data region and the log region; **c** The comparison of the whole storage system lifetime with different configurations

number of erase operations is decreased with the same write intensity. Consequently, the hybrid storage system with the dynamic log assignment has the longest lifetime.

2.5 Chapter Summary

In this chapter, we present strategies to adopt STTRAM/PRAM in different levels of the memory hierarchy, ranging from on-chip caches to secondary storage, and propose corresponding modifications to facilitate the adoption. First, we presented a cache model for STTRAM L2 cache stacking. Even though replacing SRAM L2 cache with STTRAM can result in significant power savings, the drawback of using the STTRAM comes from the long latency and high energy associated with the write operations. Consequently, we propose to add a read-preemptive write buffer with aggressive priority policy, to mitigate the performance penalty caused by the long write latency; we also propose a hybrid STTRAM-SRAM L2 cache to reduce the number of write operations to the STTRAM so that the performance is improved and the dynamic power is reduced. Second, we study the data pattern of memory write operations and explore the frequent-value locality of PRAM memory. An architecture of frequent-value storage is proposed for PRAM memory based on the data locality. This approach can significantly reduce the write intensity to PRAM memory so that both lifetime and performance can be improved. After using the frequent-value storage architecture, the endurance of PRAM is improved, and the write energy is reduced. Third, we propose a hybrid architecture using NAND flash memory and PRAM, which makes it possible to exploit the advantages of both technologies. Our PRAM-based log region method has shown that the performance of the flash memory could be improved significantly. We also show that our method can decrease the overhead of read operations and increase the lifetime of the storage system. Furthermore, the energy consumption of both read and write operations is reduced.

References

1. Kim, C., Burger, D., Keckler, S.: An adaptive, non-uniform cache structure for wire-delay dominated on-chip caches. In: Proceedings of the 10th International Conference on Architectural Support for Programming Languages and Operating Systems (2002)
2. Thoziyoor, S., Ahn, J.H., Monchiero, M., Brockman, J.B., Jouppi, N.P.: A comprehensive memory modeling tool and its application to the design and analysis of future memory hierarchies. SIGARCH Comput. Archit. News **36**(3), 51–62 (2008). doi:http://doi.acm.org/10.1145/1394608.1382127
3. Loi, G.L., Agrawal, B., Srivastava, N., Lin, S.C., Sherwood, T., Banerjee, K.: A Thermally-Aware performance analysis of vertically integrated (3-D) processor-memory hierarchy. In: DAC '06: Proceedings of the 43rd Annual Conference on Design Automation, pp. 991–996 (2006)
4. Li, F., Nicopoulos, C., Richardson, T., Xie, Y., Narayanan, V., Kandemir, M.: Design and management of 3D chip multiprocessors using network-in-memory. In: ISCA '06: Proceedings of the 33rd, Annual International Symposium on Computer Architecture, pp. 130–141 (2006)
5. Chishti, Z., Powell, M.D., Vijaykumar, T.N.: Distance associativity for high-performance energy-efficient non-uniform cache architectures. In: MICRO 36: Proceedings of the 36th Annual IEEE/ACM International Symposium on Microarchitecture, p. 55 (2003)
6. Chishti, Z., Powell, M.D., Vijaykumar, T.N.: Optimizing replication, communication, and capacity allocation in CMPs. SIGARCH Comput. Archit. News **33**(2), 357–368 (2005)
7. Kahle, J.A., Day, M.N., Hofstee, H.P., Johns, C.R., Maeurer, T.R., Shippy, D.: Introduction to the cell multiprocessor. IBM J. Res. Dev. **49**(4/5), 589–604 (2005)
8. Kongetira, P., Aingaran, K., Olukotun, K.: Niagara: a 32-way multithreaded SPARC processor. IEEE Micro **25**(2), 21–29 (2005)
9. Magnusson, P.S., Christensson, M., Eskilson, J., Forsgren, D., Hållberg, G., Högberg, J., Larsson, F., Moestedt, A., Werner, B.: Simics: a full system simulation platform. Computer **35**(2), 50–58 (2002)
10. http://www.spec.org
11. Bienia, C., Kumar, S., Singh, J.P., Li, K.: The parsec benchmark suite: characterization and architectural implications. In: Proceedings of the 17th International Conference on Parallel Architectures and Compilation, Techniques (2008)
12. Skadron, K., Stan, M.R., Sankaranarayanan, K., Huang, W., Velusamy, S., Tarjan, D.: Temperature-aware microarchitecture: Modeling and implementation. ACM Trans. Archit. Code Optim. **1**(1), 94–125 (2004). doi:http://doi.acm.org/10.1145/980152.980157
13. Puttaswamy, K., Loh, G.H.: Thermal Herding: Microarchitecture Techniques for Controlling Hotspots in High-Performance 3D-Integrated Processors, pp. 193–204 (2007). doi:http://dx.doi.org/10.1109/HPCA.2007.346197
14. Brooks, D., Tiwari, V., Martonosi, M.: Wattch: a framework for architectural-level power analysis and optimizations. In: ISCA '00: Proceedings of the 27th Annual International Symposium on Computer Architecture, pp. 83–94. ACM, New York, NY, USA (2000). doi:http://doi.acm.org/10.1145/339647.339657
15. Black, B., Annavaram, M., Brekelbaum, N., DeVale, J., Jiang, L., Loh, G.H., McCaule, D., Morrow, P., Nelson, D.W., Pantuso, D., Reed, P., Rupley, J., Shankar, S., Shen, J., Webb, C.: Die stacking (3D) microarchitecture. In: MICRO 39: Proceedings of the 39th Annual IEEE/ACM International Symposium on Microarchitecture, pp. 469–479 (2006)
16. Kgil, T., D'Souza, S., Saidi, A., Binkert, N., Dreslinski, R., Mudge, T., Reinhardt, S., Flautner, K.: PicoServer: Using 3D stacking technology to enable a compact energy efficient chip multiprocessor. Proc. 2006 ASPLOS Conf. **41**(11), 117–128 (2006)
17. Dong, X., Wu, X., Sun, G., Xie, Y., Li, H., Chen, Y.: Circuit and microarchitecture evaluation of 3D stacking magnetic RAM (MRAM) as a universal memory replacement. In: DAC '08: Proceedings of the 45th Annual Conference on Design Automation, pp. 554–559 (2008)
18. Zhang, Y., Yang, J., Gupta, R.: Frequent value locality and value-centric data cache design. SIGPLAN Not. **35**(11), 150–159 (2000). doi:http://doi.acm.org/10.1145/356989.357003

19. Yang, J., Zhang, Y., Gupta, R.: Frequent value compression in data caches. In: Proceedings of MICRO 2000, pp. 258–265. doi:http://doi.acm.org/10.1145/360128.360154
20. Zhou, P., et al.: Frequent value compression in packet-based noc architectures. In: Proceedings of ASP-DAC 2009, pp. 13–18
21. Yang, J., Gupta, R.: Energy efficient frequent value data cache design. In: Proceedings of the MICRO 2002, pp. 197–207
22. Mehrara, M., Austin, T.: Exploiting selective placement for low-cost memory protection. ACM Trans. Archit. Code Optim. **5**(3), 1–24 (2008). doi:http://doi.acm.org/10.1145/1455650. 1455653
23. Yang, J., Gupta, R.: Frequent value locality and its applications. ACM Trans. Embed. Comput. Syst. **1**(1), 79–105 (2002). doi:http://doi.acm.org/10.1145/581888.581894
24. Kim, Y., et al.: Atlas: a scalable and high-performance scheduling algorithm for multiple memory controllers. In: Proceedings of HPCA 2010, pp. 1–12 (2010). doi:10.1109/HPCA. 2010.5416658
25. Zhou, P., Zhao, B., Yang, J., Zhang, Y.: A durable and energy efficient main memory using phase change memory technology. In: Proceedings of the 36th Annual International Symposium on Computer Architecture, pp. 14–23 (2009). doi:http://doi.acm.org/10.1145/1555754.1555759
26. Qureshi, M.K., Franceschini, M.M., Lastras-Montano, L.A.: Improving read performance of phase change memories via write cancellation and write pausing. In: Proceedings of HPCA '10, pp. 123–132 (2010)
27. Toshiba America Electronic Components, Inc.: NAND flash applications design guide (2004)
28. Lee, S., Moon, B.: Design of flash-based DBMS: an in-page logging approach. In: Proceedings of ACM International Conference on Management of Data (2007)
29. Birrel, A., Isard, M., Thacker, C., Wobber, T.: A design for high-performance flash disks. Technical Report MSR-TR-2005-176, Microsoft Research (2005)
30. Park, S., D.Jung, Kang, J., Kim, J., Lee, J.: CFLRU: a replacement algorithm for flash memory. In: Proceedings of International Conference on Compilers, Architecture and Synthesis for Embedded Systems, pp. 234–241 (2006)
31. Samsung Electronics: datasheet K9G8G08UOM (2006)
32. Samsung Electronics: datasheet KPS1215EZM (2006)
33. Lam, C.: Cell design considerations for phase change memory as a universal memory. In: Proceedings of International Symposium on VLSI Technology, Systems and Applications, pp. 132–133 (2008). doi:10.1109/VTSA.2008.4530832
34. Zhang, Y., et al.: An integrated phase change memory cell with Ge nanowire diode for cross-point memory. In: Proceedings of IEEE Symposium on VLSI Technology, pp. 98–99 (2007). doi:10.1109/VLSIT.2007.4339742
35. Lee, K., et al.: A 90nm 1.8V 512Mb diode-switch PRAM with 266MB/s read throughput. In: Proceedings of IEEE International Solid-State Circuits Conference pp. 472–616 (2007). doi:10.1109/ISSCC.2007.373499
36. Nirschl, T., et al.: Write strategies for 2 and 4-bit multi-level phase-change memory. In: Proceedings of IEEE International Electron Devices Meeting, pp. 461–464 (2007). doi:10.1109/ IEDM.2007.4418973
37. http://www.tpc.org
38. Park, Y., Lim, S., Lee, C., Park, K.: PFFS: a scalable flash memory file system for the hybrid architecture of phase-change RAM and NAND flash. In: Proceedings of ACM Symposium on Applied Computing (2008)
39. Increasing flash solid state disk reliability. Technical Report, SiliconSystems (2005)
40. Chang, Y., Hsieh, J., Kuo, T.: Endurance enhancement of flash-memory storage systems: an efficient static wear leveling design. In: Proceedings of Design Automation Conference, pp. 212–217 (2007). doi:http://doi.acm.org/10.1145/1278480.1278533
41. Jung, D., Chae, Y., Jo, H., Kim, J., Lee, J.: A group-based wear-leveling algorithm for large-capacity flash memory storage systems. In: Proceedings of International Conference on Compilers, Architecture and Synthesis for Embedded Systems, pp. 160–164 (2007). doi:http://doi. acm.org/10.1145/1289881.1289911

42. http://hammerora.sourceforge.net/
43. Shibata, N., et al.: A 70 nm 16GB 16-Level-Cell NAND flash memory. Proc. IEEE Symp. VLSI Circ. **43**(4), 929–937 (2007). doi:10.1109/JSSC.2008.917559

Chapter 3
Moguls: A Model to Explore the Memory Hierarchy for Throughput Computing

3.1 Introduction

In the last chapter, we explore the benefits of replacing SRAM/DRAM with NVMs. In this chapter, we introduce a model of designing memory hierarchy for throughput computing with different memory technologies.

Throughput computing (TC) refers to trading off latency or single-thread performance for higher overall computational throughput. Throughput computing involves performing a huge number of calculations with a large amount of parallelism. Applications span many domains and are already critical on a variety of platforms, including high-performance computing (HPC) machines (e.g., molecular dynamics simulations [1]), commercial servers (e.g., database query processing [2]), and client machines (e.g., image/video processing [3]). Consequently, throughput computing is gaining prominence, leading many architects to examine how to best design systems for such applications.

In throughput computing, much debate focuses on the best programming model and the best organization for the functional units on a throughput computing processor. However, there is little doubt that memory bandwidth is a key factor in the performance of throughput computing systems. Memory bandwidth is critical because of two main reasons:

- Throughput computing applications inherently have generous amounts of parallelism that processors can take advantage of via multi-threading and single-instruction-multiple-data (SIMD) execution; thus, hardware can consume data at high rates. Systems designed to perform well on throughput computing applications achieve high performance by exploiting their inherent parallelism. These systems support large numbers of threads and/or use wide SIMD execution (e.g., Sun's Niagara [4] and nVidia's Tesla [5]), which puts a lot of pressure on the memory system. In throughput computing, memory latency is typically not a bottleneck since the latency can be hidden via multithreading (e.g., for GPUs) or hardware prefetching (e.g., for CPUs). However, bandwidth is a potential bottleneck.

G. Sun, *Exploring Memory Hierarchy Design with Emerging Memory Technologies*, Lecture Notes in Electrical Engineering 267, DOI: 10.1007/978-3-319-00681-9_3, © Springer International Publishing Switzerland 2014

- Many throughput computing applications have inherently large working sets (e.g., tens to hundreds of MB), which are unlikely to fit in conventional on-die SRAM caches for the foreseeable future. Further, unlike more traditional workloads (e.g., those similar to TPC benchmarks), some throughput computing applications show a sharp drop in performance once caches are too small to hold their working sets— we refer to this as a performance cliff. Thus, these applications are likely to be bandwidth-bound at main memory unless some significant changes are made to the memory hierarchy. Even in the general-purpose computing community, memory bandwidth was predicted to become a performance bottleneck [6], for example, due to reduced bandwidth efficiency from overly-aggressive hardware prefetching.

Throughput computing can be conducted on either CPU-based or GPU-based systems. CPUs historically have used multi-level cache hierarchies to reduce average latency and reduce bandwidth demands on larger capacity levels. GPUs have so far relied on shallow hierarchies—they hide latency with multithreading and use much higher bandwidth in main memory than CPUs (i.e., GDDR vs. DDR). The common wisdom is that the GPU approach is better for throughput computing applications. While this may be true, GPUs are designed primarily for graphics. Consequently, neither GPUs nor CPUs have a memory hierarchy designed specifically for throughput computing with an emphasis on bandwidth improvement. There are a few simple techniques to improve bandwidth efficiency of a system [7], but they are insufficient for the large bandwidth requirements of some throughput computing applications. We also cannot rely on ever-increasing main memory bandwidth to meet the needs of throughput computing. For example, today's systems that tout good performance for throughput computing applications (e.g., nVidia's Tesla) do so by providing large main memory bandwidth via the use of GDDR rather than improving bandwidth efficiency. However, GDDR has fairly strict capacity limits and is much more power hungry than conventional DRAM modules, which reduces its desirability for throughput computing, and makes it an unfavorable choice for general-purpose systems trying to improve their throughput computing performance. Furthermore, technology trends indicate that growth in bandwidth demand outpaces growth in bandwidth supply for all DRAM-based memory. (For example, historically processor throughput has grown by a factor of 1.5x per year, while DRAM bandwidth has grown by only 1.3x per year [8]). Consequently, it is necessary to study the memory hierarchy design for throughput computing platforms with an emphasis on bandwidth improvement.

Our Contribution Given that bandwidth is the key memory system characteristic for throughput computing, architects would like to optimize the memory hierarchy design with bandwidth improvements as the first priority design goal and try to find out (1) the number of levels in the optimal cache hierarchy, (2) the capacity and bandwidth of each level, and (3) the appropriate memory technology (SRAM, eDRAM, or other emerging memory technologies) of each level. To help computer architects quickly explore the design space for memory hierarchies in throughput computing platforms, this chapter makes the following contributions.

- *An analytical model called Moguls is proposed to quickly estimate the performance for applications on specific memory hierarchy designs.* The model is based on the *bandwidth demand* of an application and the *bandwidth provided* by the memory hierarchy design. The bandwidth demand/provided is defined at all memory capacities, and is described as a *capacity-bandwidth (CB)* curve (Sect. 3.2). The CB curve can facilitate a quick estimation of whether the bandwidth provided by the memory hierarchy can satisfy the bandwidth demand, and guide design improvements (for example, estimating the impact of changing the capacity of existing levels of memory, or of adding extra levels of memory).
- *The usefulness and the effectiveness of the model is demonstrated by exploring the memory hierarchy design for multi-programmed workloads running on a high throughput processor.* First-order approximations are proposed to help us quantitatively determine: (a) the most energy-efficient capacity and bandwidth of the levels in the cache hierarchy, (b) the optimal number of cache levels, (c) potential performance benefits of multiple levels of memory assuming a fixed power budget. That is, given a power budget, our model and theories suggest the best memory hierarchy (in terms of capacity and bandwidth provided). We also explore how to choose the optimal levels, capacity, and bandwidth when multiple memory technologies are provided. In order to validate our theories, we find the optimal memory hierarchy design with exhaustive simulations for real cases. We show that our theories match experimental results.

3.2 Moguls Memory Model

Designing a memory hierarchy to maximize performance and minimize power consumption is very challenging. How many levels of cache should we have? What capacities should they have? What bandwidths should they provide? On which memory technology should they based? To help solve this complex optimization problem, we first propose a way to analytically model a memory system's specifications and an application's memory requirements. We later show how to use this model to design an optimal memory hierarchy.

3.2.1 Problem Description

Figure 3.1 shows a computing system with multiple cores and a multi-level memory hierarchy. The hierarchy has n levels of shared cache (M_1 to M_n). Each level of cache can provide bandwidth BP_i to the next higher level of cache. Main memory provides bandwidth B_M. The system runs an application with a peak instruction throughput of T. This results in an aggregate bandwidth requirement from the cores, BR_C. Each level of cache filters out some of the requests, reducing the bandwidth requirement out of level i to (BR_i). In order to actually achieve a throughput of T, the bandwidth

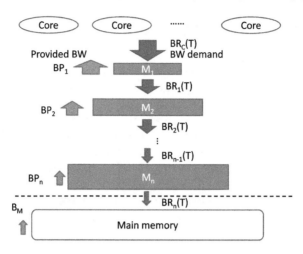

Fig. 3.1 Bandwidth requirements of the memory hierarchy

requirement of each level must be met by the level below:

$$BR_C(T) \leq BP_1; \quad BR_i(T) \leq BP_{i+1} \ (1 \leq i < n); \quad BR_n(T) \leq B_M \tag{3.1}$$

If the bandwidth requirement out of a level is greater than the bandwidth provided by the level below, the application is bandwidth-bound and the achieved throughput (i.e., performance) is below the system's peak.

Similar reasoning applies for systems with private caches or caches shared by only a subset of cores—the bandwidth requirement of each core (or subset of cores) must be separately satisfied to see peak throughput. Consequently, we discuss only shared caches, but our conclusions can be extended to cover other types of cache hierarchies.

3.2.2 Moguls Memory Model

We now use the concepts from Fig. 3.2 to construct a model, called *Moguls*, to reason about throughput computing memory systems. The Moguls model maps a computing system and an application's characteristics to capacity-bandwidth (CB) coordinates, as shown in Fig. 3.2. For each point in CB-space, the *y*-axis represents the required/provided bandwidth of a cache level, and the *x*-axis represents the capacity of the cache level. In the Moguls model, two *CB curves* are used to estimate the throughput of an application running on a system: (1) the system's provided CB curve, (2) the application's demand CB curve.

The *provided CB curve* is defined by the capacity and the effective bandwidth provided by each level of a system's memory hierarchy, and typically forms a

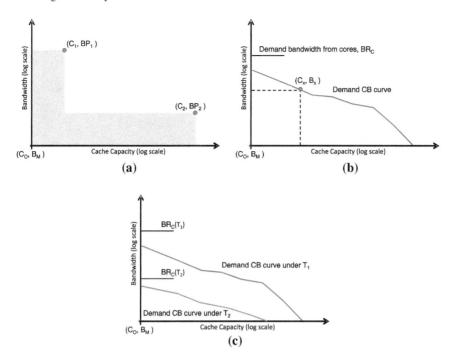

Fig. 3.2 CB curves from the Moguls model: **a** the provided CB curve of a system; **b** the demand CB curve of an application; **c** *two* demand CB curves of the same application at different computing throughputs

"stair-case" shape. The bandwidth a system provides at a given capacity is determined by the level of the memory hierarchy with that capacity, or the next largest capacity if no level exists with that capacity. Figure 3.2a shows an example of a provided CB curve (the boundary of the shaded portion) for a system with two levels of cache—one is relatively small but with high bandwidth, and the other is relatively large but with low bandwidth. Note that the origin of the coordinate system is the point (C_O, B_M), where C_O represents the minimum capacity of a cache design that is available, and B_M is the bandwidth provided by main memory.

The *demand CB curve* represents an application's capacity and bandwidth requirements—if those requirements are met, the application sees the same performance it would on a memory system with infinite capacity and bandwidth. The demand curve is defined by an application's working set sizes and the rate of execution of loads and stores. It can be derived from what is commonly referred to as a working set plot, which shows cache miss rate versus cache capacity. The working set plot is then scaled to factor in the rate that the application's loads and stores are executed in the underlying hardware (assuming infinite bandwidth provided). Figure 3.2b gives an example demand CB curve. The point (C_x, B_x) on the curve represents: *given a cache with capacity C_x, the required bandwidth to the next level down in the memory hierarchy is B_x.* The bandwidth requirement of the cores, $BR_C(T)$, is shown in the

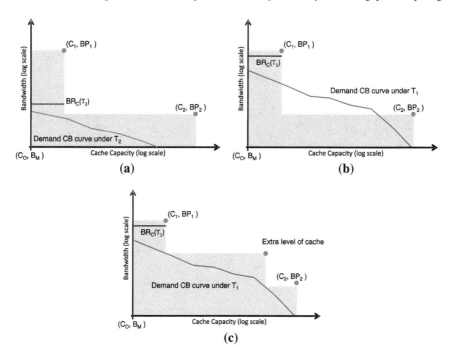

Fig. 3.3 **a** A demand CB curve is satisfied by a provided CB curve; **b** a demand CB curve is NOT satisfied by a provided CB curve; **c** a demand CB curve is satisfied by a provided CB curve by adding an extra level of cache

figure on the y axis. Note that since the scale is log-log, the y-axis is *not* at a cache capacity of zero; thus, BR_C is slightly above the y-intercept of the demand CB curve. $BR_C(T)$ is a function of the cores' throughput, as previously described. In Fig. 3.2c, two demand CB curves are shown for the same application for different computing throughputs—a larger computing throughput naturally results in a higher bandwidth requirement.

Figure 3.3 shows how we combine the provided and demand CB curves to determine whether the application bandwidth requirements are satisfied. In Fig. 3.3a, the provided CB curve is strictly above the demand CB curve—each level of the memory hierarchy can satisfy the bandwidth requirement from the level above, as defined in *Equation (1)*. In Fig. 3.3b, however, the provided CB curve dips below the demand CB curve. Consequently, the second level cache cannot satisfy the bandwidth requirement from the first level cache; thus, the system fails to achieve throughput T_1 with the example two-level cache design.

In the example, we can modify the two level hierarchy in three ways to satisfy *Equation (1)*: (1) increase the bandwidth of the larger, lower bandwidth level, (2) increase the capacity of the smaller, higher bandwidth level, or (3) add one or more new levels to the hierarchy with bandwidths and capacities between the two existing levels. Figure 3.3c illustrates the latter option.

We call our model the Moguls model because if we view the provided CB curve as a ski slope, a hierarchy with many levels has many bumps (corners), known as moguls. A memory hierarchy with smaller bumps (i.e., more graceful bandwidth degradation) is likely to provide higher performance, just as a ski slope with smaller bumps allows for faster skiing.

3.2.3 Generation of Provided CB curve

We now describe how to generate a provided CB curve to match a given demand CB curve (i.e., how to design a cache hierarchy to just meet an application's bandwidth requirements). The cache capacities are chosen arbitrarily here; later, we will address how to choose optimal capacities.

- **Step 1** The first level cache must provide bandwidth of at least $BR_C(T)$ to satisfy the cores' requirements. As shown in Fig. 3.4, we choose the point (C_1, BP_1) for the first level cache. The point (C_1, BP_1) represents a cache level with capacity C_1 and can provide bandwidth BP_1.
- **Step 2** After choosing the capacity and bandwidth for a level, we draw a vertical line from the point representing the cache till it hits the demand CB curve. The y coordinate of this intersection point is the bandwidth requirement of that cache; thus, this defines the bandwidth provided for the next level. For example, the point (C_2, BP_2) in Fig. 3.4 is chosen as the next level. Note that the corner point (C_1, BP_2) is on the demand CB curve.
- **Step 3** Step 2 is repeated till the bandwidth requirement is no more than B_M. The provided CB curve is obtained by connecting these points together, as shown in Fig. 3.4.

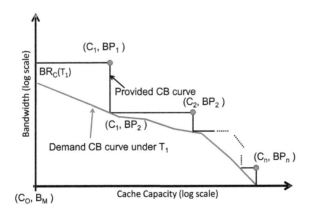

Fig. 3.4 Generating a provided CB curve from a given demand CB curve

3.3 Designing a Memory Hierarchy with Moguls

In this section, we show how to apply the Moguls model to help design a memory hierarchy optimized for power and performance. The model itself is a tool to help us analyze behavior of a specific, well-understood workload for different points in the memory system design space. To design an energy-efficient memory hierarchy for a large workload domain, we first augment the model with some assumptions about (a) the workload characteristics, and (b) power consumption of memory.

3.3.1 Approximations Used to Apply the Moguls Model

- **Approximation-1** *The demand CB curve is represented as a straight line with slope $-\frac{1}{2}$ in log-log space.*

Since it may be impractical to collect the detailed memory requirements of even one workload of interest (not to mention a whole workload domain), we approximate them. We represent target workloads with a single demand CB curve that follows the so-called 2-to-$\sqrt{2}$ rule [9]. This rule of thumb says that for complex workloads, cache miss rate varies with cache size according to an inverse power law, where the power is -0.5—if cache size is doubled, miss rate drops by a factor of $\sqrt{2}$. Studies have shown that the power actually lies between -0.3 and -0.7 [9]. Figure 3.5a illustrates the cache miss rates of several multi-programmed workloads for various cache capacities. An idealized 2-to-$\sqrt{2}$ curve (the dashed line) is also shown for comparison. Since the data is plotted on a log-log scale, the 2-to-$\sqrt{2}$ curve is a straight line with slope $-\frac{1}{2}$. This line is a reasonable first-order approximation of the measured CB curves.

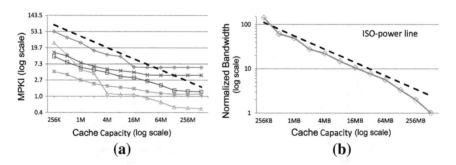

Fig. 3.5 First-order approximations. **a** The cache miss rates per thousand instructions (MPKI) for various multi-programmed workloads under different cache capacities. **b** An iso-power line for SRAM caches (45 nm process technology)

In reality, across all workloads, we expect to see a diverse set of demand CB curves. Therefore, this approximation will be much better for some workloads than others. We address the accuracy of this approximation in Sect. 3.5.

- **Approximation-2** *The access power of a cache is approximately* $\rho\sqrt{Capacity} \times Bandwidth$.

To quantitatively reason about energy efficiency, we must introduce a relationship between power, bandwidth, and capacity to the Moguls model. We use a first-order approximation that access power consumption to a cache level is proportional to the square root of its capacity times its bandwidth, i.e., $power \propto \sqrt{Capacity} \times Bandwidth$. The rationale for this approximation is that the power consumed in transporting data from/to memory cells dominates the total power of a cache, especially for large capacities. Therefore, two factors affect the power consumption: the data transfer rate and the transfer distance from/to the memory cells. The data transfer rate is proportional to the bandwidth, while the distance the data travels is proportional to the physical dimension of the memory array, i.e., the square root of capacity.

ρ is a constant determined by the process technology and some features of the cache design such as number of ports and number of banks. In the rest of this section, we assume ρ is constant, and the same for all levels of cache. We relax this assumption in later sections.

Figure 3.5b shows the results of this approximation. We use the well-known cache simulator CACTI [10] to collect data on a set of cache configurations with different capacities and bandwidths, but the same power. These data are represented with the so-called *iso-power curve* in CB coordinates. This iso-power curve closely matches the reference dashed line with slope $-\frac{1}{2}$, showing that this approximation is reasonable. The most deviation occurs in the tail region of the iso-power line, when the cache capacity is larger than 128 MB. This is an effect of leakage power, which decreases the slope of the iso-power line as cache capacity increases. Leakage power's impact depends on its fraction of the total power, and this increases with capacity. However, the results in Sect. 3.5 show that, for throughput computing systems, the high bandwidth usage promises that dynamic power consumption will dominate. For SRAM caches smaller than 128 MB, leakage power consumption has little impact on the quality of our iso-power line approximation. In addition, we introduce some memory technologies with relatively low leakage power in Sect. 3.4. Including these memory technologies as options when designing the cache hierarchy makes our approximation more accurate for caches with large capacities.

3.3.2 Cache Hierarchy Optimized for Energy-Efficiency

Using the two approximations, we can use the Moguls model to analytically derive the most energy-efficient memory hierarchy (number of cache levels and the capacity and bandwidth for each) for a given core bandwidth requirement (BR_C).

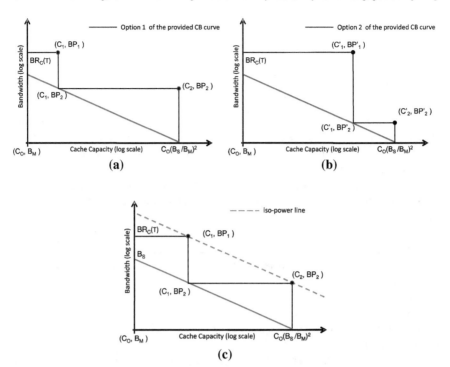

Fig. 3.6 a First option of the provided CB curve based on the same demand CB curve. **b** Second option of the provided CB curve based on the same demand CB curve. **c** An optimized cache hierarchy. The two levels sit on an iso-power line

Figure 3.6a, b give an example demand CB curve defined by $BR_C(T)$ and the 2-to-$\sqrt{2}$ rule. They also show two different provided CB curves that meet the bandwidth requirements. Both have two levels: one is defined by (C_1, BP_1) and (C_2, BP_2), and the other by (C'_1, BP'_1) and (C'_2, BP'_2). If a larger capacity first level cache $(C'_1 > C_1)$ is chosen, the bandwidth requirement to the second level cache can be reduced $(BP'_2 < BP_2)$. Although both designs provide the same throughput, they may have different power consumption.

Let us assume a two-level hierarchy, as in Fig. 3.6c, defined by (C_1, BP_1) and (C_2, BP_2). The total power consumption of these two levels is

$$P = \rho\sqrt{C_1}BP_1 + \rho\sqrt{C_2}BP_2 \tag{3.2}$$

The y-intercept of the demand CB curve is (C_O, B_S), and the x-intercept is $(C_O(\frac{B_S}{B_M})^2, B_M)$ because the slope is -0.5. The following equations are obtained based on the Moguls model,

$$BP_1 = BR_C(T); \quad C_1 = C_O\left(\frac{B_S}{BP_2}\right)^2; \quad C_2 = C_O\left(\frac{B_S}{B_M}\right)^2 \qquad (3.3)$$

If we substitute these into Eq. (3.2), we get

$$P = \rho\sqrt{C_O}\frac{B_S}{BP_2}BR_C(T) + \rho\sqrt{C_O}\frac{B_S}{B_M}BP_2 \qquad (3.4)$$

To find the BP_2 that minimize the power consumption, we need to solve $\frac{dP}{dBP_2} = 0$. We find

$$BP_2 = \sqrt{BR_C(T)B_M} \qquad (3.5)$$

A key property is that the provided bandwidth of the second level cache is the *geometric* midpoint between $BR_C(T)$ and B_M, the bandwidth requirement of the cores and the provided bandwidth of main memory. Furthermore, both levels of the cache consume the same amount of power as $\rho\sqrt{C_O}B_S\sqrt{\frac{BR_C(T)}{B_M}}$. That is, the two levels of cache are on an iso-power line (i.e., have the same power consumption).

The results for a minimum power consumption of a two-level hierarchy can be extended to n levels of cache. Specifically, for an n-level cache design, each level has identical power consumption in order to minimize total power consumption. In other words, all points representing the caches are on the same iso-power line. Due to page constraints, we describe the induction proof verbally, as follows:

1. Assume we have n-levels of caches.
2. Assume the second to the nth levels will be on an iso power line, but we have the flexibility to give more power or less power to the first-level cache.
3. When we give more power to the first-level cache (say, higher capacity), we can use less power for the rest of the levels to satisfy the throughput constraints. Alternatively, when we give less power to the first-level cache, we have to use more power for the rest of the levels.
4. In order to minimize the overall power, using a similar derivation to that shown earlier, we find that the first-level cache should also be on the same iso power line with the rest of the levels.

Thus, we prove our point that every level must on the same iso-power line. Furthermore, the bandwidth and capacity of each level will be evenly distributed in log-log space. The overall power consumption will be

$$P = n\rho\sqrt{C_O}B_S\sqrt[n]{\frac{BR_C(T)}{B_M}} \qquad (3.6)$$

Based on the results of minimum power of n levels, we can also calculate the optimal levels to further improve the energy-efficiency, by setting $\frac{dP}{dn}$ to zero. The optimal number of levels $n_{opt} = \ln\frac{BR_C(T)}{B_M}$. Note that $B_{ratio} = \frac{BR_C(T)}{B_M}$ represents the ratio of the cores' bandwidth requirement to memory's provided bandwidth. This means that as the bandwidth gap between the cores and main memory increases, more levels of cache are required to achieve the most energy-efficient cache hierarchy. This matches the historical trend, which has seen the number of levels of cache increase over time.

3.3.3 Throughput with Power Consumption Budget

We now add a power budget constraint to our optimization problem. While we've shown how to determine the most energy-efficient cache hierarchy that satisfies a given throughput requirement, this hierarchy may consume more than the allowed power budget. In this case, any memory hierarchy within the budget will have lower throughput than desired. The $BR_C(T)$ and the demand CB curve scale approximately linearly with throughput—as instruction throughput reduces, loads and stores execute at a lower rate. Thus, a throughput reduction results in a parallel downward shift of the demand CB curve in log-log space. Let the throughput reduction and the power budget be α and P_B, respectively. The following equation should be satisfied, in which T_{real} is the realized throughput.

$$P_B = n\rho\sqrt{C_O}(\alpha B_S)\sqrt[n]{\frac{\alpha BR_C(T)}{B_M}};$$

$$BR_C(T_{real}) = \alpha BR_C(T); \qquad T_{real} = \alpha T \qquad (3.7)$$

3.3.4 Throughput with Peak Bandwidth Constraint

In the previous discussion, we assumed that for a given ρ, we can build a cache with any capacity and bandwidth. In reality, the design space may be limited. Figure 3.7a shows a Moguls model plot with an extra line labeled "peak-bandwidth curve". This indicates a maximum possible bandwidth at each capacity. In the example, the most energy-efficient cache hierarchy has two levels of cache. However, the provided bandwidths of both caches exceed the peak bandwidth constraint. Thus, this optimal hierarchy cannot be built.

Figure 3.7b shows how we can provide the desired throughput by adding an extra level of cache. The extra level increases the power budget, reducing the energy efficiency of the memory system. If the power budget is violated, we must reduce the throughput. Figure 3.7c shows a sub-optimal two-level hierarchy that is within the power budget and doesn't violate the peak bandwidth constraint.

Alternatively, we can change the peak bandwidth constraint by altering the cache design or moving to a different process technology (i.e., by altering ρ). For example, the peak bandwidth of a cache can be improved by adding an access port. However, the area of the cache is greatly increased with the extra port. This indirectly increases

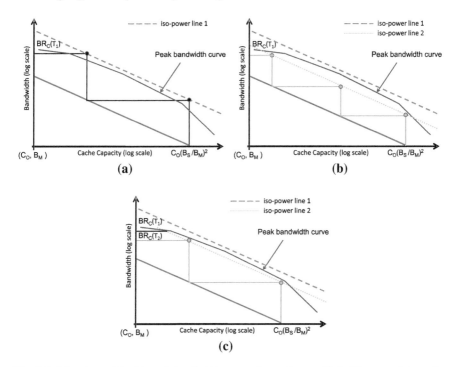

Fig. 3.7 Moguls model with peak bandwidth constraint: **a** the provided CB curve violates the constraint; **b** the provided CB curve satisfies the constraint by adding an extra cache level; **c** the provided CB curve satisfies the constraint by degrading throughput

the access energy, increasing ρ, and therefore the power consumption of the hierarchy. Thus, there is a trade-off between the peak bandwidth constraint and energy efficiency. When designing a memory hierarchy, one could feed characteristics of different cache designs into the Moguls model.

3.4 Memory Hierarchy Design with Hybrid Technologies

The factor ρ has a big impact on the optimal memory hierarchy. We've shown that for a fixed cache organization and process technology, we can approximate ρ as a fixed parameter in the Moguls model. However, in reality ρ may not be constant over the full range of capacities and bandwidths of interest. This is one reason why real memory hierarchies use multiple levels composed of different memory technologies. For example, the last level cache of Power7 is composed of embedded DRAM (eDRAM) [11]. Compared to a traditional SRAM cache, eDRAM has a slower access speed, higher density, and lower power consumption. This means that eDRAM is more energy-efficient when used as a large capacity, lower bandwidth cache—in other words, when used as a lower level cache. Recently, more emerging

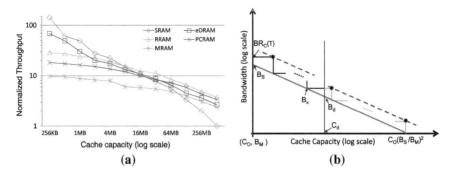

Fig. 3.8 **a** Iso-power lines for *five* different memory technologies. All points consume the same power, even across technologies. **b** The Moguls model extended to hybrid memory technologies. The *vertical line* indicates the crossover point for the iso-power lines

memory technologies such as STTRAM, RRAM, and PRAM have been proposed as potential candidates for future memory hierarchy designs [12–14]. Hybrid systems employing one or more of these technologies may provide the best solution, so we consider them here.

Figure 3.8a shows iso-power lines for various memory technologies. While we form lines by connecting iso-power points from the same memory technology, *all* points on the graph represent configurations with the same power consumption. One interesting observation is that the iso-power lines cross each other. For example, iso-power lines for SRAM and STTRAM cross at the point where cache capacity is about 64 MB. It means that SRAM has a lower ρ than STTRAM when cache capacity is less than 64 MB, but STTRAM has a lower ρ when the cache capacity is larger than 64 MB. Because it takes more energy to access an STTRAM cell than to access an SRAM cell, the power consumption of an SRAM cache is lower than that of an STTRAM cache for small capacities. However, as capacity increases, the energy consumed at the wire connections becomes more important. Because STTRAM has a much higher density than SRAM, the wire connections of an STTRAM cache are much smaller than that of an SRAM cache, and thus consume less power. Consequently, when the capacity is large enough, the smaller energy consumption of wire connections for STTRAM offsets the overhead of higher energy consumption of cell accesses and STTRAM caches have a lower ρ.

The cross points of iso-power lines of different memory technologies depend not only on the process technologies and cache organization, but also on workload behavior. For example, the write access energies for PRAM and STTRAM are much higher than their read access energies. For the same access intensity to a cache, if the percentage of writes is higher, the energy-efficiency of PRAM and STTRAM is lower. On the contrary, for memory technologies such as SRAM and eDRAM, write operations burn a similar amount of energy as read operations. Consequently, iso-power lines for different memory technologies will cross at different points in

CB space for different workloads. For the example shown in Fig. 3.8a, about 10 % of the operations are writes.

In order to explore the benefits of using hybrid memory technologies in memory hierarchy design, we extend the Moguls model to support hybrid technologies, as shown in Fig. 3.8b. From the conclusion of the single level memory technology, the minimum power consumption of the left-hand-side technology and right-hand-side technology are

$$P_{min1} = n_1 \rho_1 \sqrt{C_O} B_S \sqrt[n_1]{\frac{B R_C(T)}{B_x}};$$

$$P_{min2} = n_2 \rho_2 \sqrt{C_d} B_d \sqrt[n_2]{\frac{B_x}{B_M}}, \tag{3.8}$$

where the left-hand-side technology has n_1 levels, the right-hand-side technology has n_2 levels, and $\sqrt{C_O} B_S = \sqrt{C_d} B_d$. The total power of the memory hierarchy is $P = P_{min1} + P_{min2}$. The goal is to find the values for n_1, n_2, and B_x that minimize overall power consumption. The procedure to solve this system is as follows:

1 Find B_x such that $\frac{dP}{dB_x} = 0$, assuming n_1 and n_2 are fixed.
2 Find n_1 such that $\frac{dP}{dn_1} = 0$, assuming $n = n_1 + n_2$.
3 Find n, where $n = n_1 + n_2$, such that $\frac{dP}{dn} = 0$, by iterating the two steps above.

Intuitively, when the capacity cross-over point C_d is closer to the right hand side, n_1 tends to be larger than n_2. Conversely, when the cross-over point is closer to the left-hand-side, n_1 tends to be smaller than n_2. This is because we should implement the larger number of levels using the technology that is more efficient over a larger part of CB space.

3.5 Experiments and Validation

We now validate the Moguls model in two sets of experiments. (1) We use Moguls to derive the most energy-efficient cache hierarchy for a workload, and compare that design to others via a set of exhaustive simulations. (2) We use Moguls to derive the highest throughput cache hierarchy under a fixed power budget, and compare that to the throughput of other iso-power designs nearby in the design space, again via simulation. For most workloads we evaluate, Moguls helps us derive the optimal cache hierarchy (or very close to it). In addition, we quantify the throughput improvement achieved by using multiple memory technologies.

3.5.1 Experimental Setup

We collect the parameters of different cache configurations from a version of CACTI that we extended to support multiple memory technologies. We also extend the cache model with a mode to process requests in a pipelined manner to estimate the peak bandwidth requirement. The detailed cache configurations are listed in Table 3.1. For our experiments, we focus on validating our model and methodology for a subset of the memory hierarchy between current on-die caches and main memory. We seed the memory hierarchy with a 256 KB cache that can meet the cores' bandwidth requirements at that capacity. In our experiments, we consider additional levels of cache between this one and main memory, but do not alter the 256 KB cache in any way. Therefore, the "core bandwidth requirement" we feed to the model is actually the bandwidth requirement of this 256 KB cache. The other levels of cache are determined either by the Moguls model or are predetermined by us (e.g., when we do an exhaustive search). In practice, every additional level of cache carries overhead such as a communication interface and buffers. Consequently, the total number of cache levels should be limited. In this work, we consider designs with no more than three additional levels (i.e., besides the 256 KB cache).

We use the ZESTO [15] simulator to measure performance. It is configured to model an eight-core processor. Each core is similar to Intel's Core i7. The simulator captures data addresses from all loads, stores, and prefetch operations. We use this information to calculate the memory access intensity, and use that to compute the energy consumption of the cache hierarchy. We scale the frequency of the cores to control the processor's total peak instruction throughput (i.e., assuming it is not bandwidth bound). To study memory hierarchies for future computing systems, we evaluate processors with total instruction throughput of 32, 64, 128, and 256 billion instructions per second. The average bandwidth provided by main memory is assumed to be 8 GB/s [16].

For our first set of experiments, we explore all reasonable points in the entire cache hierarchy design space (exhaustive simulations). That is, we consider all possible hierarchies in terms of number of levels (from 0–3 levels) and capacity of each level (256 KB–512 MB), although we restrict capacities to powers-of-two, and require that capacity increases as we move down the hierarchy. In these experiments, we choose the bandwidth for each cache to just meet the bandwidth requirements of the workload being evaluated. In our second set of experiments, we use the configuration determined by the Moguls model as a starting point, and evaluate a set of cache hierarchies that modify that configuration in certain ways (e.g., have twice the cache capacity). For those experiments, we carefully choose the cache bandwidths to keep all of the hierarchies at a fixed power budget. For both sets of experiments, we simulate the workloads on all candidate sets of cache hierarchies, including the Moguls-derived hierarchy, and report performance and/or power statistics.

Since we focus on a subset of the memory hierarchy close to main memory, we select throughput computing workloads that are bandwidth-bound at main memory on modern systems. Our workloads are sets of multiprogrammed benchmarks from

Table 3.1 Cache
configurations and design
space

Read/Write port:1, 64B Cache line, 4-Way	
Cache process technology	45 nm
Cache rapacity range	512 KB–512 MB
Memory technologies	SRAM, eDRAM, STTRAM
	PRAM, RRAM
Level one cache	256 KB (fixed)
Possible extra cache level	0–3
Cache policy	Inclusive

SPEC2006 and PARSEC [17]. We randomly choose benchmarks from the full set to help us create a diverse set of demand CB curves.

Our goal has two folds: (1) to validate the Moguls model itself, and (2) to validate our methodology for applying the Moguls model to cache hierarchy design, as described in Sect. 3.3. Therefore, for our experiments when applying the Moguls model, we assume that we do not have the actual demand CB curves for our workloads, and must approximate them. Our methodology approximates the shape of a demand CB curve using the 2-to-$\sqrt{2}$ rule. We first demonstrate how well this fits our workloads.

We examined the real demand CB curves of hundreds of workloads and identified the most common patterns (i.e., shapes of the curves). Figure 3.9 shows examples of the two most common patterns. We also show the approximated demand CB curves (straight dash lines) for comparison. We show cache misses per thousand instructions (MPKI) versus capacity curves—these can be converted to bandwidth versus capacity by multiplying the MPKI by the instruction throughput and cache line size (a constant factor). The two patterns shown cover more than 90 % of the workloads we examined. In pattern-1, the demand CB curve is relatively close to the line generated from the 2-to-$\sqrt{2}$ rule. In pattern-2, the real demand CB curve follows the 2-to-$\sqrt{2}$ rule quite well for low capacities. However, the MPKI levels off at some point, often due to streaming data (i.e., data accessed only once, and so not captured in even an infinite capacity cache). This pattern normally exists in workloads that have higher MPKIs, compared to those with pattern-1. The impact of different patterns on the accuracy of the Moguls model is further discussed later with experimental results. We defer discussion of workloads that fall outside these two patterns to later.

To estimate a demand CB curve, we need more than the shape of the curve. While the 2-to-$\sqrt{2}$ rule provides a first order approximation of the shape (a straight line with slope -0.5 in log-log space), we still need a starting point (e.g., y-intercept) for the curve. Further, since this exercise is intended to mimic a real design process, we should limit ourselves to a simple process to obtain the starting point. In this case study, we use a fixed 256 KB cache in the memory hierarchy, and seek to design the levels between that cache and memory. Therefore, we can use the bandwidth requirement of the 256 KB cache as our starting point. In other words, we need only measure the MPKI out of the 256 KB cache, and then apply the 2-to-$\sqrt{2}$ rule,

Table 3.2 Optimized cache levels: Moguls estimation versus Simulations

Instruction throughput	Pattern-1		Pattern-2	
	Moguls	Simulation	Moguls	Simulation
32 billion/s	0	0	1	1
64 billion/s	1	1	2	2
128 billion/s	1	1	2	2
256 billion/s	2	2	3	3

to estimate a workload's demand CB curve. This simple method is reasonable for designing future systems, especially when the memory design space is very large and it is not feasible to collect the real demand CB curves through simulation. Our results will validate that this simple approximation method is sufficient for our case study. We discuss other modeling methods in the next section.

3.5.2 Validation of Moguls Model

Our first set of experiments validates that **when optimizing for energy efficiency, the Moguls model provides the best configuration**. All configurations considered meet the throughput requirements of the processor (from 32 to 256 billion instructions per second), and thus give the same performance. However, they consume different amounts of power.

We start by looking at the number of levels in the best hierarchy versus the hierarchy derived from Moguls. For Moguls, this number is derived from B_{ratio} (see Sect. 3.3.B), or the ratio of core bandwidth required to main memory bandwidth provided. Table 3.2 shows the estimated best number of levels (labeled "Moguls") and the best number found through our exhaustive simulations (labeled "Simulation"). The table shows results from a representative workload (shown in Fig. 3.9) from each of the two key patterns of real demand CB curves.

The Moguls estimates exactly match the simulation results for the representative workloads because their real demand CB curves are close enough to the approximation we use. Further, we evaluated hundreds of workloads in this manner, and for most that fit the two key demand CB curve patterns, the results calculated by the Moguls model match well with those obtained from simulation. Overall, 92% of the workloads have an exact match between the Moguls estimate and simulation results.

We can make some additional observations. As computing throughput increases, the number of levels in the most energy-efficient hierarchy increases. This matches our expectation: increased compute throughput means a larger bandwidth gap between the core's requirements and what main memory provides; consequently, more cache levels should be added to the cache hierarchy in order to maximize energy efficiency. Similarly, the representative workload from pattern-2 requires more levels than the workload from pattern-1 because it has a larger MPKI (i.e., higher bandwidth requirement).

We next validate that the cache capacities selected by the Moguls model match those from the optimal cache hierarchy. Table 3.3a shows the detailed cache configurations for the same representative workloads. The Moguls model assumes that cache capacity is a continuous quantity. Therefore, when applying it to real system design, we need to round the cache capacities that it generates to the closest available buildable capacity. In our experiments, we assume that the capacity of cache is limited to integer powers of two.[1] Consequently, we round the Moguls-generated capacities to the closest integer power of two (shown in the parenthesis) before comparing to the simulation results.

For the workload with pattern-1, the cache capacity results exactly match the simulation results. This is because the demand CB curve of the workload is a good fit to the Moguls approximation. For the workload with pattern-2, the cache capacities calculated by the model match the simulation results for instruction throughputs of 32, 64, and 128 billion instructions per second, but do not match at the highest instruction throughput, 256 billion instructions per second. The reason can be found

Table 3.3 Capacity of each cache level: Moguls estimation versus Simulation results

(a) Experimental results of workload for pattern-1

Throughput (Billion/s)	256		128		64		32	
Cache level	Moguls	Sim.	Moguls	Sim.	Moguls	Sim.	Moguls	Sim.
Level 1(MB)	1.6 (2)	2	2.8 (2)	2	0.77 (1)	1	0	0
Level 2(MB)	10.4 (8)	8	0	0	0	0	0	0

(b) Experimental results of workload for pattern-2

Throughput (Billion/s)	256		128		64		32	
Cache level	Moguls	Sim.	Moguls	Sim.	Moguls	Sim.	Moguls	Sim.
Level 1(MB)	1.04 (1)	2	1.7 (2)	2	0.84 (1)	1	0.70 (1)	1
Level 2(MB)	4.49(4)	16	12.7(16)	16	3.3(4)	4	0	0
Level 3(MB)	19.6(16)	32	0	0	0	0	0	0

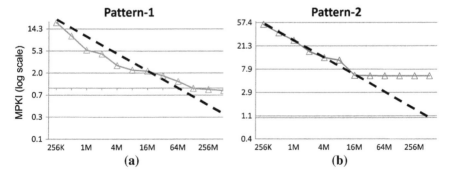

Fig. 3.9 *Two* primary patterns of demanded CB curves and the approximation in Moguls. MPKI (y-axis) versus Capacity (x-axis)

[1] A real designer may have more flexibility in capacities he/she can choose, and can pick something closer to what Moguls generates.

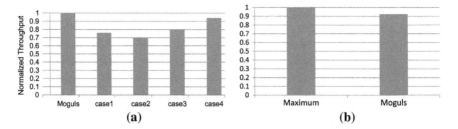

Fig. 3.10 a The throughput under a power consumption budget (the example workload with pattern-2); **b** Average throughput achieved using the Moguls configuration normalized to the target throughput

Table 3.4 Cache capacities of different configurations

	Moguls	Case1	Case2	Case3	Case4
Level 1(MB)	M1	0.5*M1	M1	2*M1	M1
Level 2(MB)	M2	M2	0.5*M2	M2	2* M2

in the tail region of the demand CB curve shown in Fig. 3.9b. The cache miss rate flattens out at a capacity of 16 MB, creating a tail to the demand CB curve. For lower throughput requirements, the model still generates the optimal hierarchy because it does not attempt to pick a cache with a capacity in the tail region. However, for very high throughput requirements, the optimal hierarchy has more levels, and one of them is in the tail region. For all workloads in our study, the Moguls model achieves about 80 % accuracy of calculating both optimal cache level and optimal cache capacity.

Our second set of experiments validates that when designing under a fixed power budget, Moguls chooses the cache hierarchy with highest performance. All cache configurations considered have the same power consumption.

We use the following procedure to generate and evaluate the various cache configurations. First, for each workload, the cache hierarchy derived from the Moguls model is adopted in the simulator. Second, we measure the maximum throughput of this cache hierarchy under the chosen power budget through simulation. Third, the cache hierarchy is replaced with different configurations. These configurations have their capacities based on the Moguls configuration, as explained below. We evaluate all possible configurations to find the maximum computing throughput for caches with those capacities under the power budget. Note that the power consumption budget needs to be carefully chosen for each workload so that the computing throughput is controlled in a reasonable range. We find the minimum power consumption required to achieve a throughput of 128 billion instructions per second for each workload, and this minimum power consumption is set as the power consumption budget.

Figure 3.10a compares the computing throughput (normalized) of the workload in Fig. 3.9b for different cache configurations. The first result (labeled with "Moguls") uses the cache hierarchy derived from the Moguls model. The other cache configurations are modifications of the Moguls hierarchy. Table 3.4 gives the details. For our two representative workloads, the Moguls-derived configurations are the most energy efficient. Therefore, the other configurations are all lower performance—in some cases, they have much lower performance.

In Fig. 3.10b, the average results of computing throughput (normalized) are compared for all workloads. The first bar is the maximum computing throughput that can be achieved. Recall that we perform exhaustive simulations for each workload to determine the minimum power needed to achieve 128 billion instructions per second. Thus, our "target" throughput is 128 billion instructions per second. The second bar in the graph shows the average throughput across all workloads when using the Moguls-derived hierarchy under the power budget. On average, the Moguls-derived hierarchies achieve 92.3 % of the maximum computing throughput.

3.5.3 The Analysis of 2-to-$\sqrt{2}$ Approximation

Although the 2-to-$\sqrt{2}$ approximation for demand CB curves fits most of our workloads, for about 8 % of them, this approximation is not a good fit. These workloads fit a third pattern of demand CB curve shape. Figure 3.11a shows three example workloads that fit this pattern. As can be seen in the figure, this shape is not very close to a line with slope −0.5. Therefore, the demand CB curve estimated via the method in Sect. 3.5.1 has significant error.

This shape of demand CB curve has a distinct "knee". The MPKI does not decrease much as we add cache capacity until we hit the knee. This pattern is very common in applications with a single, dominant working set—until the cache can hold the entire working set, there is little benefit to increasing capacity, and once the entire working set *does* fit, there is little benefit to any further increases in capacity. Multi-programmed workloads can also exhibit this shape of demand CB curve when their component applications have similar working set sizes. For example, if we run eight copies of the same application with the same input on a processor, it is highly possible that the real demand CB curve fits pattern-3. For a demand CB curve with this shape, we often cannot derive the optimal cache hierarchy from the Moguls model. The differences between the Moguls-derived hierarchy and the optimal hierarchy are caused by the knee. If a cache level is chosen with a capacity close to the knee, the real bandwidth requirement may be much larger than that calculated by the approximated demand CB curve. This is the source of most of the differences between the Moguls results and the best results from exhaustive simulations.

When designing real systems, we will typically not target a single workload so that the error of our approximation is mitigated. We demonstrate this experimentally as follows. Assume we want to design a system that will run all three workloads in Fig. 3.11a, and we want a single demand CB curve to represent all three. The figure shows the real demand CB curve for each of the three workloads. We choose our estimated demand CB curve to be a 2-to-$\sqrt{2}$ line starting from the geometric mean of the MPKIs of the three workloads at 256 KB. The figure shows this as a dashed line. We then use this demand CB curve and the Moguls model to derive a cache hierarchy.

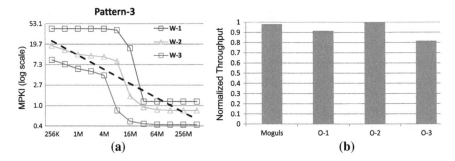

Fig. 3.11 **a** The *third* pattern of demand CB curves. **b** Relative average throughput for different cache configurations

Table 3.5 Cache configurations

	Moguls	O−1	O−2	O−3
Level 1(MB)	2M	512K	1M	1M
Level 2(MB)	16M	32M	16M	8M

In Fig. 3.11b, we compare the average throughput achieved by different cache configurations across all three workloads, under a power consumption constraint. The detailed cache configurations are listed in Table 3.5. The label "Moguls" indicates the cache hierarchy derived from the Moguls model. "O-1", "O-2", and "O-3" represent the cache hierarchies optimized for each of the three workloads, respectively, as determined through exhaustive simulation. Despite approximating these demand CB curves with a straight line, the Moguls hierarchy is within 3 % of the best hierarchy, "O-2".

3.5.4 Improvements from Hybrid Memory Technologies

As discussed earlier, the energy efficiency of the memory hierarchy may be improved by using multiple memory technologies. In this subsection, we evaluate the benefits of using a hybrid hierarchy. As shown in Fig. 3.8a, the iso-power lines of these memory technologies cross at different points in CB space, which result in different benefits.

For example. iso-power lines of SRAM and eDRAM cross at the point where the cache capacity is about 16 MB. This suggests that caches larger than 16 MB should be implemented with eDRAM to improve power efficiency. Using this 16 MB threshold, we find that across all of our workloads, 17 % of the individual caches levels should be implemented with eDRAM. We re-perform exhaustive simulations on our workloads (using the fixed power budget from earlier in this section), allowing for cache hierarchies using both SRAM and eDRAM caches. The average throughput improves by 5 % when eDRAM is available.

Using the similar method, we evaluate the combination of using SRAM with STTRAM/RRAM/PRAM. The results show that the throughput can be improved by 11 % when RRAM is available. For STTRAM and PRAM, the improvement of throughput is less than 3 %. Due to the high write access energy, the caches implemented from these two technologies are more energy-efficient than those from SRAM only when the cache capacity is very large (>64 MB for most workloads). In the future, as working set sizes increase with larger inputs, we believe that the benefits for hybrid memory hierarchies will also increase since many of the alternative memory technologies are energy efficient for large caches, but not small ones.

3.6 Future Work

We believe that the Moguls model can serve as a universal model to accelerate the memory hierarchy design process for future throughput computing systems. We demonstrate that using some first-order approximations, the model can help designers quickly narrow down the huge set of design choices they are faced with when beginning their work. While our evaluation focuses on caches, we believe the model can be applied to other parts of the storage hierarchy, such as main memory or even hard disk drives.

There are some additional issues that, if addressed, would make the model even more valuable.

Cost: Adding levels to a cache hierarchy may increase the number of physical components in a system (e.g., number of die or physical packages). This has a direct impact on manufacturing cost. It would be valuable to model not only a power budget for a system, but a cost budget as well.

I/O Power: Our model assumes that the relative power of different memories is determined solely by capacity and bandwidth. We ignore I/O power for off-chip memories, in part because there is so much variance among the options (e.g., die stacking vs. front-side bus).

Cache Policies: Multi-level cache hierarchies can enforce an inclusion property, an exclusion property, or something in between. Our model assumes inclusion between all levels, which greatly simplifies some of the math involved. Most (but not all) current systems enforce inclusion between all levels in the hierarchy. However, this wastes some capacity compared to an exclusion property. Also, our model assumes that all memories have the same cache line size and associativity and that all use an LRU replacement policy; alternative policies may result in slightly different provided CB curves and iso-power lines.

3.7 Chapter Summary

The bandwidth gap between processing cores and off-chip memory is increasing, especially with the emergence of throughput computing systems. Traditional cache hierarchies with only a couple levels cannot bridge the gap efficiently. In addition, constraints such as a power budget aggravate the problem. Thus, memory hierarchy design should be re-thought for future throughput computing systems. We propose a model to estimate performance of particular designs, and show how with some first-order approximations, we can use the model to quickly estimate the optimal point in a complex design space that includes multiple memory technologies.

References

1. Phillips, J.: Case study: molecular dynamics. In: Supercomputing 2007 Tutorial on High Performance Computing with CUDA (2007)
2. Hardavellas, N., Pandis, I., Johnson, R., Mancheril, N., Ailamaki, A., Falsafi, B.: Database servers on chip multiprocessors: limitations and opportunities. In: Proceedings of the Biennial Conference on Innovative Data, Systems Research, pp. 79–87 (2007)
3. Smelyanskiy, M., Holmes, D., Chhugani, J., Larson, A., Carmean, D.M., Hanson, D., Dubey, P., Augustine, K., Kim, D., Kyker, A., Lee, V.W., Nguyen, A.D., Seiler, L., Robb, R.: Mapping high-fidelity volume rendering for medical imaging to CPU, GPU and many-core architectures. IEEE Trans. Visual Comput. Graphics **15**(6), 1563–1570 (2009)
4. Kongetira, P., Aingaran, K., Olukotun, K.: Niagara: A 32-way multithreaded SPARC processor. IEEE Micro **25**(2), 21–29 (2005)
5. Nvidia: Tesla C1060 datasheet. http://www.nvidia.com/docs (2008)
6. Burger, D., Goodman, J.R., Kägi, A.: Memory bandwidth limitations of future microprocessors. In: Proceedings of the International Symposium on Computer Architecture, pp. 78–89 (1996)
7. Rogers, B.M., Krishna, A., Bell, G.B., Vu, K., Jiang, X., Solihin, Y.: Scaling the bandwidth wall: challenges in and avenues for CMP scaling. In: Proceedings of the International Symposium on Computer Architecture, pp. 371–382 (2009)
8. Patterson, D.A.: Latency lags bandwith. Commun. ACM **47**(10), 71–75 (2004)
9. Hartstein, A., Srinivasan, V., Puzak, T.R., Emma, P.G.: Cache miss behavior: is it sqrt(2)? In: Proceedings of the Conference on Computing Frontiers, pp. 313–320 (2006)
10. Thoziyoor, S., Muralimanohar, N., Ahn, J.H., Jouppi, N.P.: CACTI 5.1 technical report HPL-2008-20. HP Labs (2008)
11. Ware, M., Rajamani, K., Floyd, M., Brock, B., Rubio, J.C., Rawson, F., Carter, J.B.: Architecting for power management: the IBM™POWER7™approach. In: Proceedings of the International Symposium on High-Performance Computer Architecutre, pp. 1–12 (2010)
12. Wu, X., Li, J., Zhang, L., Speight, E., Rajamony, R., Xie, Y.: Hybrid cache architecture with disparate memory technologies. In: Proceedings of the 36th Annual International Symposium on Computer Architecture, pp. 34–45 (2009). doi:http://doi.acm.org/10.1145/1555754.1555761
13. Sun, G., Dong, X., Xie, Y., Li, J., Chen, Y.: A novel architecture of the 3D stacked MRAM L2 cache for cmps. In: High Performance Computer Architecture, 2009. HPCA 2009. IEEE 15th International Symposium on, pp. 239–249 (2009). doi:10.1109/HPCA.2009.4798259
14. Lewis, D.L., Lee, H.H.S.: Architectural evaluation of 3d stacked rram caches. In: Proceedings of 3DIC (September) (2009)
15. Loh, G., Subramaniam, S., Xie, Y.: Zesto: a cycle-level simulator for highly detailed microarchitecture exploration. In: Proceedings of the International Symposium on Performance Analysis of Systems and Software, pp. 53–64 (2009)
16. Ddr3 sdram standard. In: http://www.jedec.org/standards-documents/docs/jesd-79-3d (July)
17. Bienia, C., Kumar, S., Singh, J.P., Li, K.: The parsec benchmark suite: characterization and architectural implications. In: Proceedings of the 17th International Conference on Parallel Architectures and Compilation Techniques (2008)

Chapter 4
Exploring the Vulnerability of CMPs to Soft Errors with 3D Stacked Non-Volatile Memory

4.1 Introduction

Due to the continuously reduced feature size, supply voltage, and increased on-chip density, modern microprocessors are projected to be more susceptible to soft error strikes [1–5]. Consequently, the majority of the on-chip memory components (such as SRAM based structures) face exacerbating challenges. As soft error rates continue to grow traditional protection techniques such as ECC show short comings, especially in multi-bit error cases. In recent years non-volatile memory technologies, such as STT-RAM, have emerged as candidates for future universal memory. The prior work on NVM mainly focuses on the density, power, and non-volatility advantages [6–10]. In order to explore performance advantages several approaches have been proposed to use NVMs as the replacement of DRAM for the main memory [6–8], or as the replacement of SRAM for on-chip last-level caches (LLCs) [9]. Ipek et al. propose the "resistive computation", which explores STT-RAM based on-chip memory and combinational logic in processors to avoid the power wall [10].

Yet the main focus has been on its density, power advantages as well as non-volatility, the advantage of NVM's immunity to soft error strikes, however, is not yet well studied at the architectural level. Since STT-RAM storage does not rely on an electrical charge, the state of its basic storage block is not altered by an emissive particle. Recent research show that the soft error rate of STT-RAM, caused by particle strikes, is several orders lower than that of SRAM. [11–13]. Sun et al. proposed a error-resilient L1 Cache using STT-RAM [13]. The work, however, only focuses on L1 caches in a single core processor. The impact of using STT-RAM caches on the reliability of the whole cache hierarchy in a multi-core system is not studied.

While NVMs present the aforementioned advantages, it is difficult to integrate most NVM components together with traditional CMOS, due to technology compatibility issues. However, the inherent heterogeneity in 3D integration can make it possible to integrate any NVM layer in the same stack with CMOS-based processor layers efficiently. In this work, we leverage the advantages of 3D integration and NVM to improve the vulnerability of CMPs to soft errors. In particular, we focus

G. Sun, *Exploring Memory Hierarchy Design with Emerging Memory Technologies*,
Lecture Notes in Electrical Engineering 267, DOI: 10.1007/978-3-319-00681-9_4,
© Springer International Publishing Switzerland 2014

this work on inherent SER and endurance advantages of STT-RAM based caches. We explore replacing various levels of on-chip memory with 3D stacked STT-RAM to improve the soft-error vulnerability. The contributions of this work are as follows:

- We quantitatively model the vulnerability of STT-RAM to various soft errors and compare it to traditional memory technology such as SRAM.
- We utilize the low access latency through layers in 3D integration, and propose different configurations of L2/L3 caches with SRAM, eDRAM and STT-RAM. We compare these configurations, in respect of performance, power consumption, and reliability, to explore the benefits of using STT-RAM.
- To examine any performance and/or inherent soft error mitigation benefits of using STT-RAM in the processor core, we replace memory components in pipelines with STT-RAM and stack them on top of other logic parts in a separate layer. We also propose techniques to mitigate the overhead of STT-RAM's long write latency. In order to further eliminate soft errors in logic components of cores, we duplicate the core in the saved area by using STT-RAM, which verify the output of instruction execution and recover soft errors.
- We define and use a metric/method for evaluating soft error rate (SER) that evaluates vulnerability together with performance.
- We analyze the thermal characteristics of the resulting stacked configurations to indicate that the temperature profiles are within manageable ranges.

4.2 Preliminaries

In this section, we provide a brief overview of ECC protection. We also discuss the background of STT-RAM and its immunity to soft errors.

4.2.1 ECC Protection

To improve the SER vulnerability for modern processors, the on-chip memories (which are traditionally implemented with SRAM) are protected with either ECC or parity check circuit. Both of these protection schemes incur extra performance and power overhead. For example, L2/L3 caches are normally protected by ECC and L1 caches/TLBs are protected by parity checking code. Memory components in pipelines, such as register files and ROB, are often protected by parity check code or may not be protected by any mechanism at all, because they are in the critical path of the pipeline and cannot afford the performance overhead of error protection.

The most common ECC protection scheme uses Hamming [14] or Hsiao [15] codes that provide single-bit error correction and double-bit error detection (SECDED). A common solution is to apply error detecting codes (EDC) and ECC for all cache lines. Consequently, for every read or write operation, error detection/correction decoding and encoding are required, respectively.

In addition to the performance/energy overhead due to encoding/decoding, ECC protection incurs area overhead. The relative ECC overhead decreases as the size of protected block data increases. The encoding of ECC, however, are normally implemented in the granularity of a word in modern microprocessors [5, 16]. In this work, we assume that the ECC is encoded with the granularity of data word. Thus, for the 32-bit and 64-bit word sizes, the area overhead of ECC is about 22 and 13 % [4].

Previous research has presented various approaches at the architecture level to reduce the area and energy overhead of ECC protection [4, 5]. Kim proposed a hybrid protection mechanism, which applies ECC protection only to dirty cache lines, and other clean cache lines are protected by using simple parity checking codes [5]. Yoon, et al., introduced a Memory Mapped ECC [4], in which the ECC for LLCs is maintained in the main memory, and can be loaded into the LLC as normal data when necessary. Kim et al. presented a two dimensional encoding for embedded memory to detect multi-bit errors with low overhead [17].

Some of these approaches is to reduce the size of redundant ECC in the on-chip memory. The others try to improve the strength of protection using more complicated mechanism. These methods, however, either modify the memory architecture or induce extra hardware. For methods using off-chip ECC, performance is also degraded because of the extra traffic to the main memory, especially when the access intensities to L2/L3 caches are very high. We find that the on-chip ECC cannot be eliminated as far as the SRAM is still employed as the on-chip memory. The limitations of these methods will become severer as the size of SRAM caches increases and process technology scales. In addition, the encoding and decoding of ECC for error detection also consume extra energy.

4.2.2 Soft Errors of STT-RAM

Traditionally, the soft error vulnerability of a memory cell can be quantitatively expressed with the probability of state switching under a specified environment. We show that, under well controlled process, the switching probability of STT-RAM is much lower than that of SRAM/DRAM. Consequently, STT-RAM is considered to be immune to soft errors. Note that we use 65nm process technology for experiments in this section.

When a particle strikes the transistor, the accumulated charge generates a pulse of current, which may cause the switching of state in tradition SRAM/DRAM. The strength and duration of the pulse depend on the energy of particle. Prior research has shown the distribution of particle energy observed under different altitudes [18]. With the spice simulations, we observe that the amplitude of current caused by particle is much lower that of switching the state of a MTJ. More importantly, the switching of MTJ requires that the current should be kept for a long enough duration, which is at least several nano-seconds [19]. The duration of current pulse generated by a particle strike is too short to switch a MTJ. *Therefore, even if the energy of a particle*

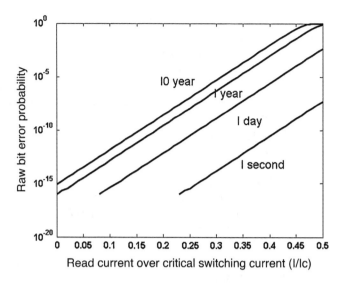

Fig. 4.1 The raw bit error probability of STT-RAM. Thermal stability: 75

strike is high enough to enable the transistor in the cell, the current cannot change the status of the MTJ.

Different from traditional memory technologies, magnetization switching of MTJ is also affected by the thermal stability. In other words, soft errors may caused by thermal fluctuation in STT-RAM. Based on prior research [20, 21], we model and simulate the switching probability under thermal fluctuation. The simulation results are shown in Fig. 4.1. The error probability is shown for different simulation duration (from 1 second to 10 years) under working temperature. The parameter *I/Ic* represents the relative reading current magnitude. *Ic* represents the magnitude of switching current of a write operation. It shows the error probability when the memory is read with different current. The 0 current represents the case that the memory cell is in the stand-by mode without being read. Although the cell is more vulnerable during the read operation, we can focus on the stand-by mode, which is the common status of memory bits in most of time. The thermal stability factor of STT-RAM is set to 75 in the experiments. The thermal stability factor of STT-RAM is a character related to different parameters including, transistor size, material and geometric ratio of MTJ, etc [21], which can be controlled under specified processing technology. From the figure, we can find that the switching probability of a STT-RAM cell under thermal fluctuation is less than 10^{-15} in a year, which is much lower than that of a SRAM/DRAM cell caused by particle strikes. For such a low probability, the well controlled STT-RAM can be considered to be immune to soft errors caused by thermal fluctuation. Besides the simulation results, the immunity of STT-RAM to soft errors is also supported by experiments on real productions [22]. The real production of STT-RAM was tested at evaluated temperature(120–240 °C) for various times and not state changes were observed.

4.3 Architecture Modification

In this section, we first introduce the baseline configurations of our 3D CMPs. Then, we propose various replacement strategies for different levels of memory hierarchy in the CMPs.

4.3.1 Baseline Architecture

Figure 4.2 shows the baseline structure of this work. There are four cores located in the layer 1. The L2 cache is located in the layer 2, which is stacked above the core layer. The four cores share the same L2 cache controller. The L2 cache controller is connected to the L2 caches by way of the through-silicon-vias between layer 1 and layer 2. There are four more cache layers stacked over the L2 cache layer because the L3 cache is normally much larger than the L2 cache. The four cores also share the same L3 cache controller. The communication between multiple L3 cache layers and the cache controller is through a bus structure, which is also implemented with TSVs. There is a router located in each layer, which connects the cache bank to the bus. This bus structure has the advantage of short inter-connections provided by 3D integration.

In this work, all replacements follows the constraint of the same area, i.e., we need to keep the similar form factor or cover the similar on-chip real estate. We estimate the area of both SRAM and STT-RAM cache banks with our extension model of CACTI [23]. We observe that the area of the STT-RAM data cell is about 1/4 of the SRAM cell area. With the same area constraint, the capacity of STT-RAM L2 cache is increased to about 4.6 times of the SRAM one after removing the ECC code. Note that the number of tag cells are also increased as we integrate more STT-RAM cache lines.

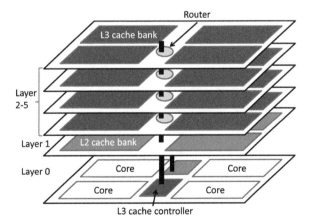

Fig. 4.2 An illustration of the proposed 3D NUCA structure

		L2	L2-ECC	L3	L3-ECC
Table 4.1 Three strategies of replacing L2/L3 SRAM caches with STT-RAM caches	(a)	STT-RAM	–	STT-RAM	STT-RAM
	(b)	SRAM	SRAM	STT-RAM	STT-RAM
	(c)	SRAM	STT-RAM	STT-RAM	STT-RAM

4.3.2 Replacing L2 and L3 Caches with STT-RAM

We propose three different replacement configurations for L2/L3 caches, which are listed in Table 4.1. The details are discussed next.

L2 STT-RAM + L3 STT-RAM

The intuitive wisdom is to replace the whole L2 and L3 SRAM caches with STT-RAM. This implies that the ECC and corresponding circuitry in both the L2 and L3 caches will be removed, adding to the potential area and capacity for more cache lines in the resulting STT-RAM caches. Similar to the previous case, both the L2 and L3 cache capacities are increased to about 4.6 times.

The first obvious advantage is that all the data cells/arrays are immune to soft error strikes. The second advantage is that the processor can have the maximum capacity of on-chip memories allowing it to achieve the lowest cache miss rates. For applications with large working set and low to moderate cache write intensities, we anticipate improved performance since the STT-RAM's limitations on write latency would be partially masked. Since the L2 cache is not the LLC, the penalty of each L2 cache miss is much reduced because of the existence of the L3 cache. The performance benefit from the lower L2 cache miss rate is reduced, compared to what we would expect in the previous configuration. Since the L1 cache is write through, for applications with intensive updates to memory, the performance could be degraded with STT-RAM L2 caches.

L2 SRAM + L3 STT-RAM

With the potential of performance degradation in the "L2 STT-RAM + L3 STT-RAM" configuration, we propose another configuration where we only replace the L3 SRAM cache with STT-RAM. Such strategy adapts the advantages from both SRAM and STT-RAM caches. We want to achieve a fast access speed from the L2 cache and get a low miss rate from the LLC (L3) cache. The L2 cache is write-back, hence the write intensity of the L3 cache would be much lower compared to that of the L2 cache. Therefore, the effect of long write latency can be effectively hidden in the L3 caches. Compared to the pure SRAM caches, the anticipated low LLC miss rate promises a general improvement of performance. With the L3 cache being the largest on-chip memory component, the raw FIT of the whole cache hierarchy is also greatly improved.

L2 SRAM + L3 STT-RAM, L2 ECC code in L3

Due to the high density of STT-RAM, the capacity of the L3 cache would be greatly increased. As we discussed earlier, 3D technology integration provides the same transparent latency for access to different layers. Therefore, we propose to implement most ECC of the L2 SRAM cache with STT-RAM and move it to the L3 cache layer to make room for enlarging the capacity of the L2 cache. A small part of ECC is kept in the L2 cache layer and implemented with SRAM. Only the ECC code of recent accessed data is stored in the SRAM part for fast access, and the rest is kept in STT-RAM. This idea is similar to that of off-chip ECC [4]. Our STT-RAM based ECC, however, induces much less overhead of performance, due to the short access latency of TSVs.

A small space in the lowest STT-RAM L3 cache layer is saved for storing the ECC of the L2 SRAM cache. We add one more set of TSVs which connect between the ECC and the L2 cache controller. Since all L2 cache lines would now be used for storing data instead of storing ECC, we expect the resulting performance to be further improved.

4.3.3 Replacing L1 Caches with STT-RAM

In modern processors, L1 caches are normally protected and monitored by parity checking codes. Such simple mechanism does not consume much area overhead, however, it can only detect soft error events but cannot correct any of them. When the L2 cache is exclusive, the data in L1 is not backuped in L1. Consequently, the L1 caches are the largest on-chip memories that may cause SEUs under particle strikes. Even if the L2 cache is inclusive, recovering data from L2 cache induces extra overhead, which can be saved by using STT-RAM L1 cache.

We propose to separate the L1 caches from the core layer and place them onto STT-RAM layers with fast access via TSVs. According to the state-of-the-art tool McPAT [24], which estimates the area of processors. In order to simplify the design, we place the L1 caches in together with L2 caches in the same layer. The L1 and L2 cache controllers can still be located in the core layer. A major objective is to keep the footprint of the processor the same. We show such a placement in Fig. 4.3.

The access intensity to L1 caches is much higher than that to L2 caches, hence, replacing SRAM L1 caches with 3D stacked STT-RAM L1 caches has an impact on the performance of CMPs. First, the access latency may increase due to traversing on the TSVs. However, prior work has shown that the latency for traversing the TSV is trivial [25]. Second, the long write latency of STT-RAM may degrade the performance of L1 caches. Prior research has shown that a SRAM buffer can help

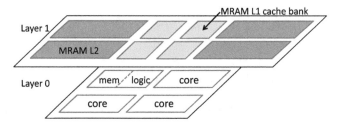

Fig. 4.3 Replace L1 caches with STT-RAM

mitigate the write overhead [13]. For the same footprint, however, the capacity of the L1 cache increases by 3x. When running applications with large working sets, the increased L1 caches can help reduce the L1 miss rates thereby improving the performance.

4.3.4 Replacing In-Pipeline Memory Components with STT-RAM

Besides caches, there are other major memory components in the processor which may be built out of SRAMs or somehow similar storage elements like register arrays, which can be affected by soft errors. These memory components are called "in-pipeline" memory components. To further explore the potential benefits of using NVM across the system storage hierarchy to improve vulnerability, we aggressively replace some of these memory components with STT-RAM in the processor.

4.3.4.1 Replacing Other Memory Components in the Processor with STT-RAM

As shown in Fig. 4.4a, the memory components in pipeline are moved to the STT-RAM layer, and a slim core without memory kept in the core layer. The impact of using STT-RAM on performance is analyzed as follows:

- Separating the memory components from the logic components and placing them into different layers may change the relative positions of these components. In the 2D case, separating memory components from the logic ones may induce timing overhead. With the 3D integration technology, it is easy to stack components, which are close to each other in original 2D placement. Finding the optimized 3D placement is out the scope of this work, we simply assume that the latency between any two component of 3D cores is kept the same as that in the 2D case.

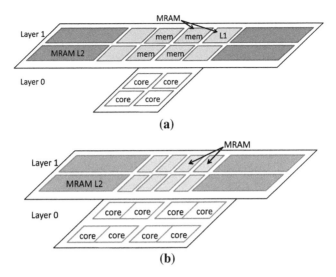

Fig. 4.4 **a** Replace in-pipeline memory with STT-RAM of same area; **b** Replace in-pipeline memory with STT-RAM of half area and duplicate the core

- For the same footprint, the capacities of these STT-RAM memory components would increase by 3x, which may potentially improve the overall system performance.
- Implementing STT-RAM structures in the pipelines would lead to longer times to fill up the pipe. We call the time to fill up the pipeline the "filling overhead".
- If the pipeline is stalled due to cache misses, it would take additional cycles to refill up the pipe after the misses resolve. Therefore, the impact of the "filling overhead" would increase. Fortunately, the large STT-RAM caches reduce the number of cache misses. However, since data dependency exists among instructions in real world workloads, the performance may be degraded in the presence of STT-RAM. For example, consider a case where an instruction in the queue consumes the result of an earlier instruction. If these two instructions are minimally far apart in the queue such that the follow-on instruction cannot simply consume via the bypass mechanism then when the first instruction completes, the result is written back to the register file before the follow-on instruction is triggered for processing. Now, if the register file is implemented with STT-RAM, the follow-on instruction could be stalled for a long while waiting for completion of the write to the register file.
- Some components in the pipeline, such as RUU and Register Files, are frequently updated. The long write latency to these components will cause the pipeline to stall leading to further performance degradation.

4.3.4.2 Techniques to Mitigate the STT-RAM Overhead

We presents techniques to mitigate the overhead caused by using STT-RAM structures in the processor pipeline. We modify the memory structures, which are frequently accessed in the pipeline, to decrease the number of stalls caused by long STT-RAM write latency.

Hybrid SRAM-STT-RAM RUU.

As the key component of an out-of-order pipeline, the RUU is most frequently updated and read. Figure 4.5a illustrates the RUU in a traditional out-of-order pipeline. "OP1" and "OP2" represent the operands of the instructions. Besides the operands data, there are several status bits in each entry. "V1" and "V2" are the validation bits of these operands. The RUU keeps snooping the buses to the Register Files, and set these bits when the operands are ready. When both operands are ready, the "R" bit will be set, which means that the instruction is ready for the ALU to execute. An RUU entry needs to be updated several times prior to instructions issuing and executing . Now, when the RUU is implemented with STT-RAM, the timing overhead of this constant updating would likely increase a lot.

To alleviate the overhead, we divide the RUU into two sub-components; the status part, which holds the status bits of each entry, and is implemented with SRAM; and the data part, which stores the corresponding operands data of each entry and is implemented with STT-RAM. With this hybrid structure, the timing overhead of updating the RUU is greatly reduced. The overall vulnerability of RUU, however, has increased because the SRAM status bits could be affected by particle strikes.

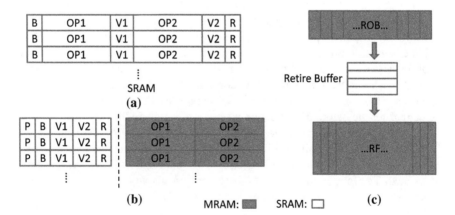

Fig. 4.5 The illustration of techniques to mitigate the overhead of using STT-RAM in the pipeline: **a** the structure of an SRAM RUU; **b** the structure of a hybrid STT-RAM-SRAM RUU; **c** the structure of retirement buffer

Since the number of status bits only occupy a small portion of the RUU, the relative increase in vulnerability is trivial, making it worth using such a hybrid architecture.

Note that the RUU is an abstract of structure, which may includes several components in the real implementation and the status bits may be different from those in Fig. 4.5a. The basic idea of the hybrid structure is to trade off the reliability for performance improvement. We would like to reduce the reliability of some components on critical timing path to mitigate the performance degradation.

Retirement Buffer

If the Register File (RF) is also implemented with STT-RAM, the time to update the RF, when retiring an instruction from the Re-Order Buffer (ROB), is also increased. As mentioned earlier, the long write latency of the STT-RAM may stall the execution of an instruction if it depends on the RF being updated. To mitigate the apparent overhead caused by the long retirement latency, we introduce a new structure called the retirement buffer, which is implemented with SRAM, as shown in Fig. 4.5c.

When an instruction is retired from the ROB, the result is stored in the SRAM retirement buffer. At the same time, write-back to the RF is initiated. When later instructions are decoded and dispatched into the RUU, they first search the retirement buffer to see whether the required operands are present. If they cannot find the operands in the buffer, they will search the RF for the required operands. The length of the buffer has an impact on the performance and vulnerability. It takes about $6ns$ to finish updating the STT-RAM RF. For a $NGHz$ processor with W retirement width, in the worst case, we need a $6NW$-entry-buffer so that the retirement speed can be competitive with an SRAM RF. In fact, since the data is normally reused in a short period, we may use less entries in the buffer to decrease the vulnerability. Due to the page limit, the impact of the buffer length is not introduced. The experiments show that a 16-entry-buffer is large enough for configurations in this work.

4.3.5 Improving Reliability of Logic

So far, our discussion has focused on the reliability of the memory hierarchy. However, the SER of the latch is catching up with the SRAM per-bit rate with a steeper slope of increase, whilst the logic combination SER is projected to increase at a much faster pace (although the absolute numbers are still smaller than SRAM or latch numbers at the present time) [3]. For silicon-on-insulator technology, going forward from 65 nm to 45 nm, the latch SER per-bit rate is predicted to increase by a factor of two to five times, and the number of latches per chip is expected to increase with integration density [3, 26].

Due to the high density advantage of STT-RAM, the replacement of STT-RAM also provides the potential of improving the reliability of components other than the memory. In previous configurations, we have discussed replacing SRAM with

STT-RAM of similar area. Alternatively, if we replace SRAM with STT-RAM of the same capacity, a lot of area can be saved. Consequently, the saved area can be utilized to improve the reliability of latches and logic components in the pipeline with different methods.

The straightforward way is to improve the reliability of logic components by increasing the size of transistors, thereby leading to hardened latches and gates. This approach, however, can only reduce the probability of SEU incidence but not fully eliminate it to zero. A more appropriate approach would be to duplicate the resulting slim core at the architectural level, as shown in Fig. 4.4b, and execute every instruction by replicating and processing it in parallel on the duplicated cores. Here again, one of two methods may be adopted. In the first method, the output of each stage in the pipeline may be compared, and whenever the two results are in disagreement, an SEU incidence may be implied. Consequently, both pipelines are flushed and the instructions are re-executed. Since this could lead to a lot more overhead, we choose the second method. In the second method, we avoid comparisons at the pipeline stages, and wait to compare the final output of the two cores. When the two results disagree, we imply an SEU, and re-execute the instruction again. This method reduces the compare logic and also avoids false positive SEUs. Whether we can duplicate these slimmed-down cores depends on the area reduction due to replacing the on-core memory components with STT-RAM. According to McPAT, memory components (including L1 caches) occupy more than 60the total area of the core. By our approach, we reduce the original area of the core by more than half after replacing all the memory components with STT-RAM.

4.3.6 Tag Array of Caches

The ECC protection is also used to protect the tag of cache, which is content address-able memory (CAM), from particle strikes. Prior work has shown that the area over-head of ECC protection in the tag array is similar to that of the data cells [27]. A lot of research has been done for the design of CAMs with the CMOS technology [27, 28], and there has been several proposed approaches for the design of STT-RAM CAM [29, 30], which show that the area of an STT-RAM tag cell is about twice of that of the CMOS tag. Implementing the tag array with STT-RAM therefore incurs an area overhead even without the ECC protection. In addition, the access energy to an STT-RAM tag array is also much larger than that to a CMOS one. Consequently, we prefer using the CMOS technology and ECC protection in the CAM-based tag array for caches with high associativity. For the SRAM tag array in L1 caches of low associativity, we can still replace them with STT-RAM.

4.4 Methodology

In this section, we present the evaluation setup. We introduce how we evaluate the vulnerability of the CMPs to soft errors. We also discuss our thermal modeling infrastructure.

4.4.1 Evaluation Setup

Our baseline configuration for our analysis is a 4-core out-of-order CMP using the Ultra SparcIII ISA. We estimate the area of the four processing cores to be about $40\,mm^2$, based on the study of industrial CMP examples [31, 32]. We assume that one cache layer fits either the 1MB SRAM or a 4MB STT-RAM cache. The configurations are detailed in Tables 4.3 and 4.4. Note that the cache access time includes the latency of cache controllers and routers. We use the Simics toolset [33] and its extension models GEMS [34] for performance simulations. We simulate multi-threaded benchmarks from the *OpenMP2001* [35] and *PARSEC* [36] suites. We pin one thread on each core during the simulation. For each simulation, we fast forward to warm up the caches, and then run ROI (region of interest [36]) code in the detailed mode.

For our 3D setup, we assume a heterogeneous 3D stack in order to incorporate the STT-RAM and SRAM layers. The device layers are assumed to be thinned to $100\,\mu m$ (with 10–$15\mu m$ for inter-layer interconnect and wiring layers) and integrated in a Face-to-Back (F2B) fashion. Final thickness of the 3D stack is similar to the starting 2D version due to the thinning. The TSV sizes are $10\mu m$ at $20\mu m$ pitch. The detailed technology parameters are listed in Table 4.2.

4.4.2 Metric for Soft Error Vulnerability

We use *mean fault per instruction* (MFPI) as the metric for vulnerability analysis, which is defined in the following equation:

$$MFPI = \frac{number\ of\ errors\ encountered}{number\ of\ committed\ instructions} \qquad (4.1)$$

Table 4.2 Technology parameters used in the experiments

Parameter	Value
TSV size/pitch	$10/20\,\mu m$
Avg. TSVs per core	<1024
Average core area	$10\,mm^2$
Silicon thickness	$100\,\mu m$ thin Si

We define *fault* to include only the errors caused by soft errors which cannot be recovered directly by the affected component. For example, the parity check code of the instruction queue can detect a single bit error but a recovery needs other fault-tolerant mechanisms (such as checkpointing/restore, which is out of scope of this book). On the contrary, the ECC code can directly recover the single bit error in the L2 caches. We do not count such self-recoverable errors as faults in this work. The MFPI does not only represent the vulnerability of the whole system but also shows the impact of each component and its contribution to the total soft error rates. In addition, it also exposes the delicate balance between performance and soft error reliability.

Though our definition of MFPI may appear similar to MITF (mean instructions to failure) as first used by Weaver et al. [37], there are differences in how we derive our values. Whereas the MITF is derived from the traditional AVF [2], which is structure-based, the MFPI is based on the instruction and tracking of the residency of its corresponding data and control signal bits as they move through the various components of the system. Our approach of deriving the errors encountered is more similar to that used by Li et al. [38], except that our focus is more instruction-based.

In this work, we trace the processing of each instruction, and calculate the time that the data of each instruction is exposed to soft error strikes. The data of an instruction may be stored in different memory components during execution. For example, the operands could be stored in the register files; some data may be stored in caches before being loaded into the pipeline. We systematically track and count all the exposed times in the various components per instruction. To achieve this, we record the time when the data is loaded into each component and the time when it is accessed by an instruction or when it is updated. Consequently, we are able to calculate the total exposed time of the data used by each instruction. If we assume that each data bit has r soft errors in a unit time when it is exposed to particle strikes, then the total number of errors that may happen in each instruction is expressed by the following equation:

$$number\ of\ errors = r \times \sum_n data\_size_i \times expose\_time_i \qquad (4.2)$$

Table 4.3 Area, access time and energy comparison of SRAM, STT-RAM, and eDRAM caches including ECC (65 nm technology) [19]

Cache size	Area (mm^2)	Read Lat. (ns)	Write Lat. (ns)	Read Ene. (nJ)	Write Ene. (nJ)	Standby power (W)
1M SRAM	36.2	2.252	2.244	1.074	0.956	1.04
4M STT-RAM	36.0	2.318	6.181	0.858	2.997	0.125
4M eDRAM	35.1	4.053	4.015	0.790	0.788	1.20

Table 4.4 Baseline configuration parameters

Processors	
Number of cores $= 4$	Frequency $= 2\,GHz$
In-order Fetch/Decode/Retire; Out-of-Order Issue/LD/ST;	
Fetch Width $=$ Decode Width $=$ Issue Width $=$ Retire Width $= 4$;	
IQ : 32 Entries, RAT and RF : 416 Entries, RUU : 32 Entries, LSQ : 128 Entries, ROB : 32 Entries	
Memory parameters	
SRAM L1 (private I/D)	16 + 16 KB, 2-way, 64B/cache line, 2-cycle, write-through
STT-RAM L1 (private I/D)	read: 2-cycle, write: 16-cycle
SRAM L2 (shared)	1 MB, 8-way, 64B/cache line, 8-cycle, write-back
STT-RAM L2 (shared)	read 8-cycle, write: 20-cycle
SRAM L3 (shared)	4 MB, 8-way, 64B/cache line, 18-cycle, write-back
STT-RAM L3 (shared)	read 18-cycle, write: 30-cycle
L1 Protection	Parity codes
L2/L3 Protection	ECC, 8B/cache line
Main memory	4 GB, 300-cycle latency

where the $data\_size_i$ represents the ith data used in the instruction and its exposed time to particle strikes is $expose\_time_i$. (Note that we assume that errors in any instruction can result in an SUE).

4.5 Experimental Results

In this section we present our experimental results on the SER vulnerability improvement, performance evaluation, power and thermal analysis of our proposed architecture configurations.

4.5.1 Performance Evaluation

The IPCs for the different configurations in Sect. 4.3.2 are compared in Fig. 4.6. As we discussed earlier, the benefit of replacing L2 SRAM caches with STT-RAM is reduced when there are L3 caches. This is because the penalty of an L2 cache miss is greatly mitigated by the L3 cache. In Fig. 4.6, the performance increases for most benchmarks when we replace both L2 and L3 caches with STT-RAM. For four of the benchmarks (*mgrid, canneal, galgel, and equake*) the performance, however, degrades with both L2 and L3 STT-RAM caches. The benefits of reducing the L2 miss rates are offset by the overhead of long write latency to the STT-RAM L2 caches. This conclusion is further supported when we only replace L3 caches with

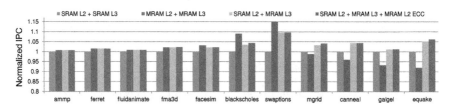

Fig. 4.6 IPC comparison between STT-RAM and SRAM for configurations of L2 and L3 caches (Normalized to the first column)

STT-RAM. The results show that the performance increases for all benchmarks. The last set of results in Fig. 4.6 shows that, after we move the ECC of L2 caches to the L3 layer, we further improve the performance due to increasing capacity in the L2 caches. The fast access speed of TSVs ensures no timing overhead for accessing the ECC in the L3 layer. Since the capacity of the STT-RAM L3 cache is large, we get more benefits by placing the ECC in the L3 layers.

In Fig. 4.7, we compare the performance of using SRAM, STT-RAM, and DRAM L1/L2 caches (Due to the page limit, we assume that L2 is the LLC. The cases with L3 caches show similar trend.) When we only replace the SRAM L2 caches, the configuration using STT-RAM L2 caches has the best performance for all benchmarks. The results of using DRAM L2 caches are even worse than using SRAM caches for some benchmarks because the DRAM has lower access speed for both read and write operations, as shown in Table 4.3. In addition, the DRAM would suffer higher access latency due to the need for constant refreshes, which we do not model in this work.

The results of replacing both the L2 and L1 caches are also shown in Fig. 4.7. Replacing the L1 has significant impact on the performance. For some benchmarks with large working sets, we get more performance benefits by increasing the capacities of the L1 caches with STT-RAM. For other benchmarks, the performance is degraded because the long write latency of the STT-RAM offsets such benefits. The results of using DRAM L1 caches show similar trend as that of using STT-RAM L1 caches. However, the performance of using DRAM L1 caches is worse than that of using STT-RAM L1 caches. This is because the L1 cache has very high read access intensity and the slow read speed of DRAM L1 caches introduces more overhead.

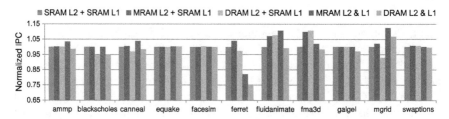

Fig. 4.7 IPC comparison of SRAM, STT-RAM and eDRAM configurations of L2 and L1 caches (Normalized to the first column)

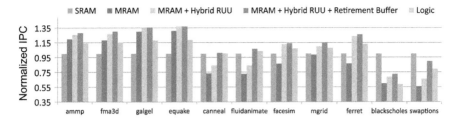

Fig. 4.8 IPC comparison of using STT-RAM and SRAM memory components in the pipeline (Normalized to the first column; "SRAM"means SRAM pipeline; "STT-RAM"means STT-RAM pipeline; "+ hybrid RUU" means using hybrid STT-RAM and SRAM RUU, "+ Retirement buffer" means using the technique of retirement buffer, "logic" means using double core)

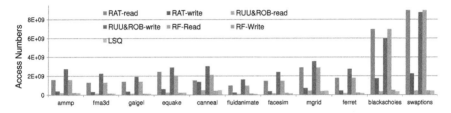

Fig. 4.9 Read and write numbers to different memory components in the pipeline

Figure 4.8 compares performance of using STT-RAM memory components in the pipeline against that of using SRAM. The IPC numbers of these two configurations are shown as the first two bars in Fig. 4.8. For the first three benchmarks (*ammp, fma3d, and galgel*), the performance increases with STT-RAM, implying that we get more benefits from increased memory capacity than the overhead of long write latency. The performance of other benchmarks, however, decrease. The second and third bars represent the IPC numbers when we employ the techniques proposed in Sect. 3.4.2. For most benchmarks, by applying those techniques, the performance improves as compared to the original pipelines using SRAM. The last two benchmarks (*blackscholes and swaptions*) remain degraded, and this is because these benchmarks have higher access intensity to the memory components in the pipeline as evidenced in Fig. 4.9, especially the write intensity to RUU and ROB. The last bar represent the case that we double the core to improve the vulnerability of logic. The performance is degraded when compared to that in the previous one because reduce the capacity of memory.

4.5.2 Soft Error Vulnerability Analysis

Figure 4.10 shows the normalized MFPI of the CMPs with different configurations of L2 and L3 caches. As we mentioned, errors recoverable in SRAM L2/L3 caches are not counted as faults when there is ECC. Hence, the vulnerabilities of our L2/L3

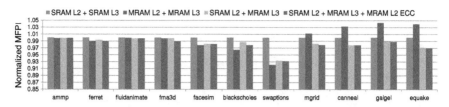

Fig. 4.10 The comparison of vulnerability between STT-RAM and SRAM for different configurations of L2 and L3 caches (normalized to the first column)

caches themselves are not affected by replacing SRAM with STT-RAM. However, the MFPIs of the L1 caches and the pipelines are related to the performance of L2/L3 caches. For most of the benchmarks, the replacement of both L2 and L3 caches reduces the period during which the data in L1 caches or pipelines are exposed to particle strikes. The MFPIs of the CMPs decrease for these benchmarks. On the other hand, the MFPIs of the last four benchmarks (*mgrid, canneal, galgel, and equake*) increase with STT-RAM L2/L3 caches.

If we compare Figs. 4.10 and 4.6, we find that the MFPI and performance are strongly correlated. When the errors of L2 and L3 caches are not counted, the MFPI of CMPs shows the opposite trend to that of IPCs. When the IPC of CMPs decrease, the period of processing data through the pipelines increases. At the same time, the period that data is exposed in the L1 caches also increases so that the MFPI increases. We can draw similar conclusions for other configurations. Consequently, when we replace the originally ECC protected caches with STT-RAM caches, we observe that the performance does not degrade while the vulnerability of the whole system improves.

The results of MFPIs for different configurations of L1 caches are shown in Fig. 4.11. Different from the cases of replacing L2/L3 caches, the errors of L1 caches are counted in the MFPI because they are normally just protected by parity checking codes. Therefore, replacing the L1 caches with STT-RAM can greatly reduce the number of MFPI by eliminating errors in the L1 caches. In addition, the higher L1 cache hit ratios can help reduce the data exposure period in the pipeline. When the performance degrades with STT-RAM L1 caches, the MFPI of the pipeline increases. The MFPI of the whole CMP, however, is still greatly reduced because the number

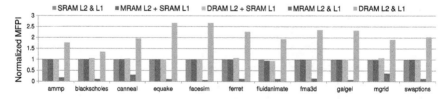

Fig. 4.11 The vulnerability comparison of SRAM, STT-RAM, and DRAM for different configurations of L2 and L1 caches (normalized to the first column)

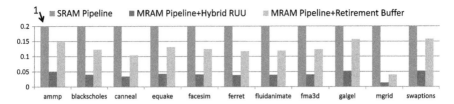

Fig. 4.12 The vulnerability comparison of using STT-RAM and SRAM memory components in the pipeline (normalized to the first column)

of errors in L1 caches is much higher than that of the pipeline. The MFPIs of using DRAM caches are also shown in the Fig. 4.11. The results show that using DRAM L1 caches increases the MFPI greatly for almost all the benchmarks. The data is exposed for a longer time to particle strikes in the DRAM L1 caches, which have larger capacity but is only protected by parity check code.

Theoretically, using STT-RAM in the pipelines can eliminate all faults in the memory components. Only the soft errors in logic components should remain, which has been shown to be relatively small at present. We do not model the soft errors in logic components in this study. By incorporating the techniques in Sect. 3.4.2 to improve the performance, we introduce extra SRAM components which may increase the MFPI. The MFPIs with using these two techniques are shown in Fig. 4.12, illustrating that we can improve the vulnerability with relatively low performance degradation. We do not show the soft errors of logic. Note that there are no soft errors in logic when we double the cores.

4.5.3 Power Consumption

As shown in Table 4.3, the STT-RAM has the advantage of low leakage power, but the write energy is higher than those of SRAM or DRAM. In this section, we quantitatively analyze the power consumption of CMPs using STT-RAM memory technologies. The power consumption of the caches and the processing cores are compared separately in order to show the impact on different components.

When we account for the energy overhead of protection mechanisms in SRAM, replacing SRAM with STT-RAM may reduce both leakage and dynamic power. In L2 or L3 caches, the power consumption of the ECC is composed of two parts. The first part comes from the access power to the extra ECC bits; and the second part is introduced by the ECC bits' encoding and decoding. For example, there are 8 Bytes of ECC codes for each 64 Bytes cache line in the caches we model. We can save about 20 % of power for each operation if we remove the ECC from the SRAM caches. Although the power of write operations increase when using STT-RAM, the total power consumption can still be reduced where the number of read operations dominates. For the parity check code, the power consumption overhead is mainly

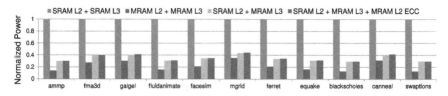

Fig. 4.13 The comparison of power consumption between STT-RAM and SRAM for different configurations of L2 and L3 caches (Normalized to the first column)

caused by the encoding and decoding operations. Our evaluation shows that removing parity check code can save about 5 % of power consumption.

The comparison of power consumption for different cache configurations are shown in Figs. 4.13 and 4.14. When there are L3 caches in the processor, the leakage power of the SRAM dominates because of the large capacity of caches. As shown in Fig. 4.13, after both L2 and L3 caches are replaced with STT-RAM, the power consumption is greatly reduced. This is because the leakage power of STT-RAM caches is much lower than that of SRAM caches. When only L3 caches are replaced with STT-RAM, the total power increases because more dynamic power is introduced in the SRAM L2, which is still lower than that of pure SRAM caches. When the ECC of the L2 cache is moved to the L3 cache layers, the leakage power of caches is kept the same. The total power, however, increases slightly because of the higher power consumption of updating STT-RAM ECC.

Figure 4.14 compares the power consumption of using SRAM, STT-RAM, and DRAM for L2 and L1 caches. The results show that the total power consumption decreases when SRAM L2 caches are replaced with STT-RAM caches. Since there are no L3 caches in the processor, the leakage power becomes less dominant. For some benchmarks, when the intensity of write operations is very high, the power consumption of using SRAM and STT-RAM caches are comparable because of large energy of writing to STT-RAM. The power consumption of using DRAM L2 caches is also shown in the figure. A DRAM memory cell has lower leakage power than that of an SRAM cell. For the similar area, however, the total leakage power (including the refresh power) of a DRAM cache is higher than that of an SRAM cache. Consequently, the total power increases with using DRAM L2 caches.

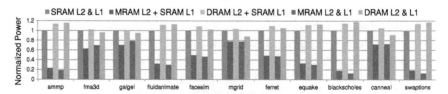

Fig. 4.14 The comparison of power consumption for different SRAM, STT-RAM, and DRAM configurations of L2 and L1 caches (Normalized to the first column)

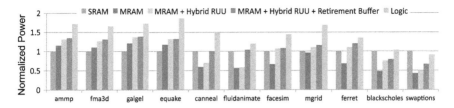

Fig. 4.15 The comparison of power consumption using STT-RAM and SRAM memory components in the pipeline (Normalized to the first column

The case is different when the L1 caches are replaced with STT-RAM. Since the capacity of the L1 cache is very small, the benefit of reducing leakage power decreases with using STT-RAM. Consequently, for benchmarks with high write intensity, the power consumption increases with using STT-RAM L1 caches. On the contrary, using DRAM L1 caches introduces more leakage power but the dynamic power decreases. The total power consumption decreases for benchmarks with high access intensity to L1 caches. For the CMPs with more processing cores, the leakage power become more dominant and the benefits of reducing leakage power increase with using STT-RAM caches. (The results are not shown due to the page limit).

The power consumption of the cores using SRAM and STT-RAM in-pipeline components are shown in Fig. 4.15. (Note that only core power, without any caches, is displayed). Although the STT-RAM write power is higher in general, some benchmarks see reduced power dissipation in the core due to the performance degradation. After using the techniques in Sect. 3.4.2, we consume more energy for accessing these extra components as shown in Fig. 4.15. We can also find that the power is greatly increased when double the cores.

4.5.4 Thermal Evaluation

We show the basic floorplan of each of the cores in our 4-core CMP in Fig. 4.16a. Figure 4.16b shows the detailed thermal map of the processor core layer. As the figure indicates, the hotspots center on the register file and the execution units. Peak temperatures have high-correlation with the processor layer power densities, since the power density of the stacked L2/L3 layers are lower in both SRAM and STT-RAM alternatives. Furthermore, as the relatively higher power processor layer is placed close to the heat sink and the SRAM/STT-RAM layers are placed closer to the board, the resulting peak temperatures are within manageable ranges.

The thermal model of our stack alternatives includes a detailed model of the device, wiring and inter-layer interconnect layers, full package, and a cooling solution. We used both ANSYS and Flotherm to model the different granularity of the stack (TSV/wiring components were modeled using ANSYS and the full-stack simulations were carried out in Flotherm). Ambient temperature is 25°C for the

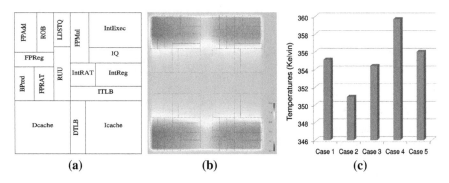

Fig. 4.16 a Basic floorplan of a core; **b** thermal map of the processor core layer (°C); **c** peak temperatures for for the thermal alternatives

simulations. We explored a range of thermal conductivity for the back-end/wiring-layers and the inter-layer interconnect layers for various 3D alternatives, and reported average values of the explored range. We assumed a cooling solution based on product specifications for blade systems with similar power ranges.

Figure 4.16c summarizes the peak temperatures for the stacking alternatives: (1) In the first configuration, the L1/L2/L3 hierarchy is fully implemented in SRAM; (2) An STT-RAM L3 is used to replace the SRAM L3; (3) L1/L2/L3 are all implemented in STT-RAM; and (4) In addition to the STT-RAM L1/L2/L3, we have pipeline units (as discussed in the previous section) replaced by STT-RAM and placed in an extra layer. Since SRAM and STT-RAM are implemented in separate layers of the stack, the functional units are moved to the corresponding device layers in the alternative implementations. As the figure shows, configuration (2) has the lower temperature than (1) (the original full SRAM stack) does, mostly due to the reduced power dissipation.

Similarly, (3) has slightly lower temperature than the original SRAM stack. However, the temperature advantages disappear in configuration (4), where the original core floorplan is changed with the STT-RAM structures are placed in another layer. As a result, the high power density functional units are placed closer to each other, and the power density is elevated in the processor layer. The peak temperatures are about 5–6 K higher in this configuration compared to (1) and (3). In configuration (5), we keep the STT-RAM replaced areas in the pipeline as white space, preserving the original core floorplan. The peak temperatures are reduced to 356 K as a result of the additional white space area. This indicates that with proper thermal planning the temperature characteristics of the alternative stacking options can all be within desired ranges. Note that the temperature is not increased much after doubling the core in case (6). It is because the distribution of power density is similar.

4.6 Chapter Summary

In this chapter, we leveraged the emerging 3D integration and STT-RAM technologies to improve the vulnerability of CMPs to soft errors. We explored various configurations where different levels of the cache hierarchy were implemented in SRAM, STT-RAM or DRAM alternatives and evaluated these alternatives with respect to soft error reliability, performance, power and temperature characteristics. Our experimental results show the trade-offs between performance and reliability using 3D stacked STT-RAM. For the average workload, replacing all levels of the memory hierarchy with STT-RAM virtually eliminates radiation induced soft errors on-chip, improves the performance by 14.5 %, and reduces power consumption by 13.44 %. The thermal characterization indicates that the resulting peak temperatures are within manageable ranges - especially with proper planning for temperatures.

References

1. Zhang, W., Li, T.: Managing multi-core soft-error reliability through utility-driven cross domain optimization. In: Proceedings of ASAP, pp. 132–137 (2008)
2. Mukherjee, S.S., Weaver, C., Emer, J., Reinhardt, S.K., Austin, T.: A systematic methodology to compute the architectural vulnerability factors for a high-performance microprocessor. In: Proceedings of MICRO, p. 29 (2003)
3. Rivers, J.A., Bose, P., Kudva, P., Wellman, J.D., Sanda, P.N., et al.: Phaser: phased methodology for modeling the system-level effects of soft errors. IBM J. Res. Dev. **52**(3), 293–306 (2008)
4. Yoon, D.H., Erez, M.: Memory mapped ECC: low-cost error protection for last level caches. In: Proceedings of ISCA, pp. 116–127 (2009). doi:http://doi.acm.org/10.1145/1555815.1555771
5. Kim, S.: Reducing area overhead for error-protecting large L2/L3 caches. IEEE Trans. Comput. **58**(3), 300–310 (2009). doi:http://dx.doi.org/10.1109/TC.2008.174
6. Zhou, P., Zhao, B., Yang, J., Zhang, Y.: A durable and energy efficient main memory using phase change memory technology. In: Proceedings of the 36th Annual International Symposium on Computer Architecture, pp. 14–23 (2009). doi:http://doi.acm.org/10.1145/1555754.1555759
7. Lee, B.C., Ipek, E., Mutlu, O., Burger, D.: Architecting phase change memory as a scalable dram alternative. In: Proceedings of the 36th Annual International Symposium on Computer Architecture, pp. 2–13 (2009). doi:http://doi.acm.org/10.1145/1555754.1555758
8. Qureshi, M.K., Srinivasan, V., Rivers, J.A.: Scalable high performance main memory system using phase-change memory technology. In: Proceedings of the 36th Annual International Symposium on Computer Architecture, pp. 24–33 (2009). doi:http://doi.acm.org/10.1145/1555754.1555760
9. Wu, X., Li, J., Zhang, L., Speight, E., Rajamony, R., Xie, Y.: Hybrid cache architecture with disparate memory technologies. In: Proceedings of the 36th Annual International Symposium on Computer Architecture, pp. 34–45 (2009). doi:http://doi.acm.org/10.1145/1555754.1555761
10. Guo, X., Ipek, E., Soyata, T.: Resistive computation: avoiding the power wall with low-leakage, stt-mram based computing. In: Proceedings of the 37th Annual International Symposium on Computer Architecture, ISCA '10, pp. 371–382. ACM, New York, NY, USA (2010). doi:http://doi.acm.org/10.1145/1815961.1816012.
11. Freescale, BRMRAMSLSCLTRL, D.N.: Freescale mram technology (2007)
12. Tehrani, S.: Status and Prospect for Mram Technology. Everspin Technologies, Inc., USA (2010)

13. Sun, H., Liu, C., Xu, W., Zhao, J., Zheng, N., Zhang, T.: Using magnetic ram to build low-power and soft error-resilient l1 cache. Very Large Scale Integration (VLSI) Systems, IEEE Transactions on PP(99), 1 (2010). doi:10.1109/TVLSI.2010.2090914
14. Hamming, R.W.: Error correcting and error detecting codes. Bell Syst. Tech. J. **29**, 14–160 (1950)
15. Hsiao, M.Y.: A class of optimal minimum odd-weight-column sec-ded codes. IBM J. Res. Dev. **14**(4), 395–401 (1970)
16. Bossen, D., Tendler, J., Reick, K.: Power4 system design for high reliability. IEEE Micro **22**(2), 16–24 (2002)
17. Kim, J., Hardavellas, N., Mai, K., Falsafi, B., Hoe, J.: Multi-bit error tolerant caches using two-dimensional error coding. In: Proceedings of MICRO, pp. 197–209 (2007). doi:http://dx.doi.org/10.1109/MICRO.2007.28
18. Mukherjee, S.: Architecture Design for Soft Errors. Amsterdam, Elsevier, Inc. (2008)
19. Gallagher, W.J., Parkin, S.S.P.: Development of the magnetic tunnel junction mram at ibm: From first junctions to a 16-mb mram demonstrator chip. IBM J. Res. Dev. **50**(1), 5–23 (2006)
20. Xiaobin, W., Yuankai, Z., Haiwen X., Dimitar, D.: Thermal fluctuation effects on spin torque induced switching: Mean and variations. J. Appl. Phys. **103**, 034507 (2008)
21. Wang, X., Chen, Y., Li, H., Dimitrov, D., Liu, H.: Spin torque random access memory down to 22 nm technology. IEEE Trans. Magn. **44**(11), 2479–2482
22. Akerman, J., Brown, P., Gajewski, D., Griswold, M., Janesky, J., Martin, M., Mekonnen, H., Nahas, J., Pietambaram, S., Slaughter, J., Tehrani, S.: Reliability of 4mbit mram. In: Reliability Physics Symposium, 2005. Proceedings. 43rd Annual. 2005 IEEE, International, pp. 163–167 (2005)
23. Thoziyoor, S., Ahn, J.H., Monchiero, M., Brockman, J.B., Jouppi, N.P.: A comprehensive memory modeling tool and its application to the design and analysis of future memory hierarchies. SIGARCH Comput. Archit. News **36**(3), 51–62 (2008). doi:http://doi.acm.org/10.1145/1394608.1382127
24. Li, S., J. H. Ahn, R.D.S., Brockman, J.B., Jouppi, D.M.T.N.P.: McPAT: an integrated power, area, and timing modeling framework for multicore and manycore architectures. In: Proceedings of MICRO (2009)
25. Loi, G.L., Agrawal, B., Srivastava, N., Lin, S.C., Sherwood, T., Banerjee, K.: A thermally-aware performance analysis of vertically integrated (3-D) processor-memory hierarchy. In: DAC '06: Proceedings of the 43rd Annual Conference on Design automation, pp. 991–996 (2006)
26. KleinOsowski, A., Cannon, E.H., Oldiges, P., Wissel, L.: Circuit design and modeling for soft errors. IBM J. Res. Dev. **52**(3), 255–263 (2008). doi:10.1109/TMAG.2008.2002386
27. Pagiamtzis, K., Azizi, N., Najm, F.: A soft-error tolerant content-addressable memory (CAM) using an error-correcting-match scheme. In: Proceedings of CICC, pp. 301–304 (2006). doi:10.1109/CICC.2006.320887
28. Pagiamtzis, K., Sheikholeslami, A.: Content-addressable memory (CAM) circuits and architectures: a tutorial and survey. IEEE J. Solid-State Circuits **41**(3), 712–727 (2006). doi:10.1109/JSSC.2005.864128
29. Wang, W., Jiang, Z.: Magnetic content addressable memory. IEEE Trans. Magn. **43**(6), 2355–2357 (2007). doi:10.1109/TMAG.2007.893305
30. Xu, W., Zhang, T., Chen, Y.: Design of spin-torque transfer magnetoresistive ram and cam/tcam with high sensing and search speed. IEEE Trans. VLSI **18**(1), 66–74 (2009)
31. Kahle, J.A., Day, M.N., Hofstee, H.P., Johns, C.R., Maeurer, T.R., Shippy, D.: Introduction to the cell multiprocessor. IBM J. Res. Dev. **49**(4/5), 589–604 (2005)
32. Kongetira, P., Aingaran, K., Olukotun, K.: Niagara: A 32-way multithreaded SPARC processor. IEEE Micro **25**(2), 21–29 (2005)
33. Magnusson, P.S., Christensson, M., Eskilson, J., Forsgren, D., Hållberg, G., Högberg, J., Larsson, F., Moestedt, A., Werner, B.: Simics: a full system simulation platform. Computer **35**(2), 50–58 (2002)

34. Martin, M.M.K., Sorin, D.J., Beckmann, B.M., Marty, M.R., Xu, M., et al.: Multifacet's general execution-driven multiprocessor simulator (gems) toolset. SIGARCH Comput. Archit. News **33**(4), 92–99 (2005). doi:http://doi.acm.org/10.1145/1105734.1105747
35. www.spec.org
36. Bienia, C., Kumar, S., Singh, J.P., Li, K.: The parsec benchmark suite: Characterization and architectural implications. In: Proceedings of the 17th International Conference on Parallel Architectures and Compilation Techniques (2008)
37. Weaver, C., Emer, J., Mukherjee, S.S., Reinhardt, S.K.: Techniques to reduce the soft error rate of a high-performance microprocessor. In: Proceedings of ISCA, p. 264 (2004)
38. Li, X., Adve, S.V., Bose, P., Rivers, J.A.: Softarch: An architecture level tool for modeling and analyzing soft errors. In: Proceedings of DSN, pp. 496–505 (2005). doi:http://dx.doi.org/10.1109/DSN.2005.88

Chapter 5
Conclusions

As the technologies scale down, the memory hierarchy implemented with traditional memory technologies cannot satisfy the requirements of high performance, low power, and high reliability. The emerging NVMs, such as STTRAM and PRAM, have the potential to replace traditional memory technologies because of their advantages. First, the high density of NVMs makes it possible to integrate more memory so that the performance can be improved. Second, since the storage cells can be powered off without losing data information because of the non-volatility of these NVMs, the standby power of the memory can be significantly reduced. Third, the NVMs are immune to radiation based soft errors so that the vulnerability of the memory hierarchy can be improved.

Although these emerging NVMs have so many advantages, it is not straightforward to directly adopt them in the memory hierarchy due to their own limitations. We have discussed that three questions need to be answered before using these NVMs. In this book, we explore the memory hierarchy design using these NVMs from different angles, and these questions are answered as following to show the scheme of adopting these NVMs efficiently:

- With specified design goals, the characters of the emerging NVMs should be considered and compared with the traditional memory technologies to decide the level of the memory hierarchy, in which the proper NVM can be adopted. For example, in order to integrated more on-chip memory, the STTRAM can be chosen to replace SRAM because STTRAM has higher density and fast read speed. In some cases, the trade-off among different design goals have to be considered comprehensively to achieve an optimized design.
- Due to limitations of these NVMs, the modifications in the architectural level are necessary to facilitate the adoption of various NVMs in the memory hierarchy. For example, the write-buffer can be improved to mitigate the problem of long write latency for STTRAM caches and the FV based storage can be used to improve the lifetime of PRAM main memory.
- Although emerging NVMs can beat traditional memory technologies in some design cases, it is still impossible to find a single memory technology, which

G. Sun, *Exploring Memory Hierarchy Design with Emerging Memory Technologies*, Lecture Notes in Electrical Engineering 267, DOI: 10.1007/978-3-319-00681-9_5, © Springer International Publishing Switzerland 2014

can be used as the universal technology in the whole memory hierarchy. On the contrary, the advantages from different memory technologies can be leveraged in together through the hybrid memory design. As shown in the book, the hybrid SRAM/STTRAM caches and hybrid PRAM/NAND-flash SSD perform better than the counterparts implemented with one pure memory technology in different level of the memory hierarchy.

Index

G. Sun, *Exploring Memory Hierarchy Design with Emerging Memory Technologies*,
Lecture Notes in Electrical Engineering 267, DOI: 10.1007/978-3-319-00681-9,
© Springer International Publishing Switzerland 2014

CPSIA information can be obtained
at www.ICGtesting.com
Printed in the USA
LVHW080149120919
630825LV00001B/3/P